THE MINEWORKERS

ROBERT DUNCAN

THE MINEWORKERS

Birlinn

First published in 2005 by
Birlinn Limited
West Newington House
10 Newington Road
Edinburgh EH9 1QS

www.birlinn.co.uk

ISBN 10: 1 84158 365 0
ISBN 13: 978 1 84158 365 5

British Library Cataloguing-in-
Publication Data
A catalogue record for this book
is available from the British Library

Designed and typeset by Mark Blackadder

Printed and bound by GraphyCems, Spain

Contents

Title page.
Pit shankers, after
completing shift at
Woolmet Colliery, 1898.
(*Collection of the Scottish
Mining Museum Trust*)

To mineworkers everywhere,
past and present

Preface

This survey interpretation of the social history of coal, lead and shale mining in Scotland would have been impossible without the work of countless people. I wish to acknowledge the contribution of fellow historians, whose research and publications were used in this study, and many mineworkers for their recorded testimony and witness. Among leading academic historians of the social history of coal mining, I wish to record particular thanks to Chris Whatley for his re-appraisal of the condition and status of mineworkers in the period of serfdom, and to Alan Campbell for his inspiring, unrivalled insight into the experience of colliery labour and coal mining communities in the nineteenth and twentieth centuries. I am also grateful to Ian MacDougall for vital coverage of the labour and oral history of coal mining and for recommending that I undertake this commission.

Thanks are due to all who helped with sources, including pursuit of the necessarily selective choice of graphics. Accordingly, I wish to acknowledge assistance from Alan Campbell; volunteers and staff at the Scottish Mining Museum, Newtongrange; Sybil Cavanagh at West Lothian Council; museums and heritage staff at North Lanarkshire Council; and the Scottish Lead Mining Museum at Wanlockhead. Guthrie Hutton's series of illustrated booklets on mining in Scotland gave useful examples of images. Staff at the National Library, Edinburgh, the Mitchell Library, Glasgow, and locally at Strathaven were courteous and helpful in dealing with many requests for a range of sources, as were secretaries in history departments at the universities of Glasgow and Strathclyde for allowing access to several doctoral theses. At Birlinn, Andrew Simmons has been a patient editor throughout the project.

Above all, on a personal note, a special thanks goes to my wife, Anne, for compiling the index, for other technical help, and for providing unfailing encouragement and support.

Robert Duncan   Strathaven, 2005

Mining and Mineworkers in Scotland before the Eighteenth Century

Origins of Commercial Mining

Coal has been extracted and mined ever since people recognised its use value for heat and fuel. In the many coal-bearing parts of Scotland, it was picked and quarried wherever a seam cropped out on the surface of the land, along river banks, and on the coastline or shore, where it gave rise to the term 'sea coal'.

The earliest recorded instances of coal-working from outcrops and shallow pits are on the monastic lands of Newbattle Abbey, Midlothian; Holyrood House at Carriden in West Lothian; Dunfermline Abbey; Culross in Fife; and on other monastic estates in the thirteenth century. The enterprising Cistercian monks at Newbattle were certainly responsible for digging holes and tunnels into the surrounding hillsides and banks of the South Esk river, and extending shallow mining operations into East Lothian around Prestongrange. They were probably the first to sell and export coal, shipping it out from small ports nearby. They were also involved in the production of salt from small coastal works, where coal was used to boil the pans of sea water, as the name suggests, at Prestonpans. The monasteries and church lands were part of the feudal system in medieval Scotland and, in the absence of firm evidence, the workforce in mining until the fourteenth century was likely to have been serf labour rather than free labour.

The earliest recorded evidence of lead mining in Scotland dates from the thirteenth century, at Crawford Muir, in remote upper Lanarkshire. In the speculative search for gold and silver, neighbouring Leadhills became the centre of extensive operations from the 1590s. That area of the Lowther hills around the adjacent mining villages of Leadhills and Wanlockhead was named 'God's Treasure House In Scotland' on account of its potential for finds of valuable minerals, although lead mining and smelting were to be the only commercially viable industries there even in modern times.

After the Reformation, monastic and church lands were broken up and divided among noble families and gentry who supported the monarchy, and from the later sixteenth century, mineral-bearing estates fell within their ownership and control. Thereafter, almost exclusively along the coastlines of the Firth of Forth, where upper coals were abundant and close to the sea, titled landowners and enterprising merchants from the burghs of Fife and the Lothians began to develop an interest in exploiting the coal measures on their properties. In Fife, where the seams were easily accessible to the available mining technique along a shore line of seventy miles as far east as Pittenweem, significant coal exporting collieries were started before the close of the sixteenth century at Culross, Dysart and Wemyss. A string of coastal collieries stretched from Kincardine to Wemyss, with Tulliallan and Torry being among the main centres of coal production throughout the seventeenth century. Alloa and Sauchie further up the Forth at Clackmannan, and Airth, on the south side, in Stirlingshire, had also become important collieries. On the south side in West Lothian, prominent collieries were already located at Grange, Bonhard, Bo'ness, Kinneil, Carriden and Linlithgow.

The scale of production and output tonnage from the Forth pits in any one year or series of years during the seventeenth century is not known, as reliable figures are not available from the archive record. During this period, the typical colliery was a very small-scale enterprise, with no more than a handful of workers at one or two producing pits. However, a major colliery such as Tulliallan may have produced around 15,000 tons in the best years, and it is reckoned that up to fifteen collieries along the Forth were able to produce up to 10,000 tons in busy years.

It is certain that the pioneering coal country between Edinburgh and Haddington in Mid and East Lothian continued overall to be the main centre of coal production in Scotland into the seventeenth century. Collieries dotted

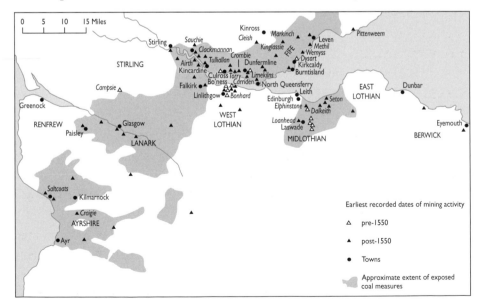

Plate 1.

The Scottish Coalfields before 1700.

(*J Hatcher*, History of the British Coal Industry, *Volume 1*)

the coast from Seton to Inveresk, while Smeaton, Pencaitland, Tranent, Elphinstone, Dalkeith and Loanhead were significant inland producers. A crop of landsale collieries to serve the local Edinburgh market had also grown up nearer the capital, on the western side of the Esk valley. There, and along the Firth of Forth, the bituminous 'great coals', with their good burning qualities, were popular fireplace fuel among commoners and kings alike. Exports of this prized coal had also for many years served a luxury market in London and in the Low Countries, although this item of southbound trade from the Fife and Lothian ports was to dwindle long before the end of the seventeenth century, to be overtaken by competition from the abundant 'big coals' of the larger Tyneside coalfield.

Early Mining Methods

Some remarkable technical achievements were made in winning coal from underground workings in the developing coalfields of east Scotland. Even so, before the eighteenth century, efforts were somewhat restricted by limited practical knowledge of mining operations. Mining methods advanced very slowly from beginnings in the thirteenth century. Accessing and working underground seams had to remain relatively simple processes, being confined to shallow depths, as the technology and expertise to withstand frequent roof collapses, flooding and gases were not yet available. Quarrying coal from outcrops was by far the simplest and most common method. Another standard method involved the more difficult and dangerous process of tunnelling to form a drift mine. The miners entered the outcrop coal at an angle and then worked under the ground along the line of the seam. Appropriately, the Scots term for such workings was an 'ingaunee' (ingoing eye).

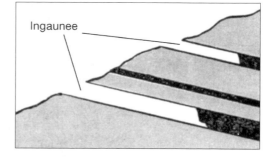

    Another established method of reaching underground seams was by sinking a short vertical shaft, and constructing a pit bottom. Here, safety conditions permitting, the working area could be extended outwards in a bell-like shape. Depending on geological conditions and technical skill, such bell pits could subsequently be linked up by tunnels, or else abandoned and other shallow pits sunk nearby in the continuing trial for workable coal.

    In their construction and function, stair pits were more sophisticated than bell pits. They were designed to reach greater depths than bell pit workings, and also to enable the coal hewers to work the seams over a longer distance than was possible at simple bell pits. The single shaft in a stair pit could be

Plate 2.

Plate 3.

A shallow bell pit.

sunk to a depth of seventy or eighty feet, and then lined with wood and clay, or with stone. Wooden platforms were added, with ladders connecting the stages at various intervals from coalface to the surface, creating an underground transport system for the work team. The more extensive underground workings would become honeycombed with 'rooms' – the workplace where the hewers dug out the coal. The method of cutting the coal around large pillars and leaving them to support the roof was, as yet, exceptional, and was not common practice until the eighteenth century.

Without doubt though, the most sophisticated advance in mining methods throughout this earlier period of development was the pit and adit system or, as it was known in Scotland, the pit and level system. This breakthrough was achieved where a drainage tunnel or level could be built below the coal-working area, to allow surplus water to escape by natural gravitation into river or sea. As a source of additional fresh air, the existence of a drainage level could also help tackle the ventilation problems that confronted coal-face workers. By the sixteenth century, successful application of this pit and level drainage method facilitated the working of coal seams up to thirty metres underground. Where drainage levels were not possible or practical, excess water in the underground approaches and workings was raised to the surface by windlass – a haulage system using buckets operated by water gins or horse gins. Although useful, such efforts were limited in their effectiveness, as they could not resolve serious drainage problems, which often led to the early abandonment of pits, and generally prevented deeper coal seams from being won or even attempted.

The coastal Culross colliery, as operated during the early seventeenth century, provides an unusual and striking example of the adoption of advanced technology. Coal was produced from the seams using the stoop and room method. The first water pumping plant in Scotland was installed here, using an endless chain with buckets. This feature of the colliery, together with the Moat Pit on a raised island at high tide and its unique two shafts – one on the shore and one on the island, with boats loading coal from the shaft on the island – were a wonder to behold and a tourist attraction of the time.[1] A visitor in 1618 described these developments:

> The mine hath two ways into it, the one by sea and the other by land
> … I went in by sea and out by land. Now men may object, how can
> a man go into a mine, the entrance of it being into the sea, but that
> the sea will follow him and so drown the mine? To which objection
> thus I answer, that at low water, the sea being ebbed away, and a
> great part of the sand bare, upon this same sand did the master of
> this great work build a round circular frame of stone, so high withall,
> that the sea at the highest flood can neither dissolve the stones so
> well compacted, or yet overflow the height of it. Within this round

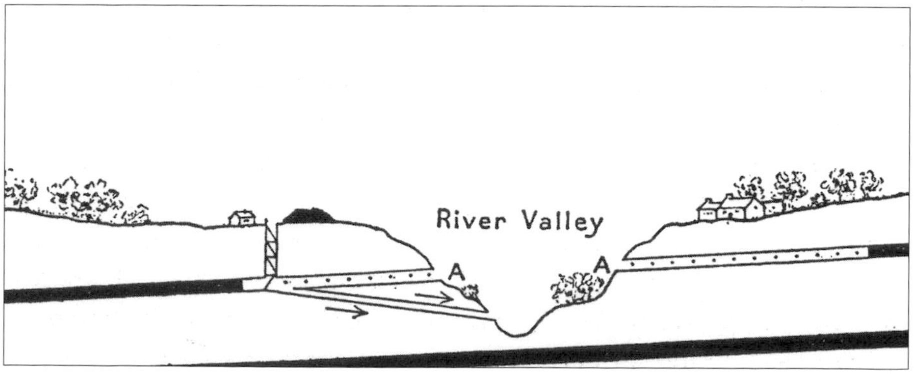

River Valley

Plate 4.

Pit and level system: worked coal shown at A. Drainage tunnel or day level shown by arrows.

frame he did send workmen to dig with mattocks, pick axes etc. They did dig forty foot down right, into and through a rock. At last they found that which they expected, which was sea cole, they following the vein of the mine, did dig forward still: so that in the space of eight and twenty years, they have digged more than an English mile under the sea. Besides, the mine is most artificially cut like an arch or a vault, all that great length, with many nooks and by-wayes: and it is so made that a man may walk upright in the most places.

The sea at certain places doth leak, or soak into the mine, which, by the industry of Sir George Bruce, is all conveyed to one near the land, where he hath a device like a horse-mill, that with three horses and a great chain of iron, going downward many fathoms, with thirty six buckets fastened to the chain; of the which eighteen go down still to be filled; and eighteen ascend up to be emptied, which do empty themselves (without any man's labour) into a trough that conveys the water into the sea again, by which means he saves his mine, which otherwise would be destroyed with the sea.[2]

This remarkable showpiece colliery with its undersea workings continued in production until the 1620s, when it finally had to succumb to the ravages of the tides. According to local tradition, when King James VI returned to Scotland in 1617, he visited Culross and asked to see Sir George Bruce's wonderful mine complex. He entered the mine at the shore end at high water time. Upon reaching the wharf on the island, he found himself surrounded by

Plate 5.
Moat Pit, Culross:
a modern sketch.

the sea. Due to the experience of previous attempts on his life, his suspicions were again aroused of a murder or abduction plot, and calmed down only when it was clear he was to be taken back onshore on a small boat already provided for the purpose. On an earlier occasion, the king had stood on Lomond Hill, above his residence at Falkland Palace, and had noticed the smoke rising from the coal-fired saltpans on the shores of the Forth. He then compared the poor and undeveloped hinterland of his kingdom of Fife and the productive mineral riches of its coastline to 'a beggar's mantle with a fringe of gold'.

Also in Fife, another notable contemporary example of capital investment, technical skill and perseverance was the drainage level driven through the rock for two miles from the shore at Methil, as ordered by the Earl of Wemyss. By the seventeenth century, the same advanced technology was applied at the lead mines of Leadhills, under owner-operator Sir James Hope, and at neighbouring Wanlockhead, on the Duke of Queensberry's estates. Long drainage levels had been driven into the hillsides, and horse gins were also used to facilitate the mining of lead ore down to twenty-four fathoms.

Pit Work: A Preview

In his account of the Moat Pit at Culross, a visitor gave us a brief glimpse of the workers as he walked along the main underground passage:

> There, young and old, with glimmering candles burning,
> Digge, delve and labour, turning and returning.

The familiar, general picture of the division of labour in underground pit work before the early nineteenth century was already being established in the larger collieries by the sixteenth century. However, this pattern of work was more common in coal workings in east and central Scotland than in the west, where the development of mining came much later. The hewer dug out the coal and the task of filling the baskets was usually completed by his wife, by another adult or by one of his children. Usually, women and girl bearers then carried the filled baskets on their backs to the foot of the shaft. In pits with stairs and ladders, bearers also had to make additional journeys up to the surface, carrying their back-breaking loads.

Before the rising demand for coal in the seventeenth century, full-time hewers and teams would have been a rarity. At most pits, they would most likely have been engaged in coal-getting on an irregular and even casual basis. They would have worked busier spells as periodic orders came in from the estate owners or outside customers, and increased seasonal demand for domestic coal ensured a busy time during November and December in readiness for the winter period. However, as discussed below, where coal pits were linked to the coastal salt-panning industry, the demand for large supplies of small coal for boiling sea-water would probably have created more year-round work for local colliers.

Coal Production: Demand and Location by 1700

By 1700, the first and foremost coal-producing region in Scotland remained concentrated along both sides of the Firth of Forth and within some of its inland parishes near the coast. Although the Forth region still occupied the primary position, many of the easily accessed high coal seams along the coast were worked out or close to exhaustion, especially those that lay above sea level and had to be drained to the shore. Nevertheless, the accessible top-quality big coal throughout the region satisfied demand from domestic consumers. The active coalfield was close to urban settlements, including the growing new burghs of the upper Forth, whose house owners increasingly used coal fires for cooking and heating. It was also intimately connected with the coastal salt-panning industry.

Although reliable figures for coal output and consumption in Scotland are not available for this period, it appears that, of an estimated 225,000 tons of total output per year in the early 1690s, over 150,000 tons was sold as household coal, with the rest going for various industrial uses. The main settlements of Haddington, Edinburgh and Leith, Linlithgow, Falkirk, Stirling, Dunfermline and Kirkcaldy were all supplied conveniently with coal by packhorse or boat from nearby pits. This was also the case with the few sizeable towns in the west of Scotland, namely Glasgow, Renfrew, Paisley and Ayr. As yet relatively small and self-contained, with around 15,000 inhabitants in 1700, Glasgow had a long-established ready access to coal supplies from small local pits. Its largest single supplier was the Gorbals colliery, south of the Clyde, forming part of the large Govan coalfield, which was to undergo rapid mining development in the eighteenth and nineteenth centuries. On the south side, in the adjacent parishes of Rutherglen and Cambuslang, the Glasgow market was also served by growing output from collieries on the Hamilton estates. Further up the Clyde valley, the inhabitants of the burghs of Hamilton and Lanark obtained coal from small local pits. In the east, Dundee and Perth were within reach of coastal shipment from the ports of the Forth, and coal was also being shipped north

Plate 6.

Stair pit.

to Aberdeen, Montrose and Arbroath. Edinburgh and Leith, with over 30,000 inhabitants in 1700, were the major domestic coal consumers in urban Scotland, providing a lucrative market for a crop of Lothian collieries. Abundant use of local coal had already contributed to the capital city's title of Auld Reekie, on account of the dense fog from thousands of smoking house-chimneys.

Of the estimated 10,000 or so tons of coal used as fuel for existing industries and land use, this was divided between lime-burning for agriculture, production of mortar for building, brewing, potteries, soap-boiling, glass-making and the traditional smith's forge. However, before the advent, in the later eighteenth century, of coal-fired furnaces for smelting in the iron industry, by far the principal industrial consumers of coal were the manufacturers of sea-salt. There were good economic reasons for integrating the two enterprises in coastal locations. Apart from cutting large coal, hewers also produced small coal and a dross-like pan-coal, which did not sell well in the open market; but it was known to be highly suited for the boiling processes in the large cast-iron salt-pans. Salt manufacturers needed six tons of coal for every ton of salt produced by the evaporation of brine. Moreover, by the end of the seventeenth century, it was evident that many collieries along the Forth were very dependent on the salt-panning industry for their economic viability. In the best years, the marine salt industry in Scotland was producing around 6,000–8,000 tons of salt per year, and consuming an estimated total of up to sixteen times that amount of coal.

The main integrated coal and salt-panning works were located along both sides of the Forth and on the east Lothian coast, but the Ayrshire coast was also a notable part of this dual enterprise. The centres of salt panning were located in Stevenston parish, at Saltcoats – where the derivation of the town's name is all too readily apparent – and nearby Ardrossan. Pan-coal was supplied from neighbouring pits.

In common with the vast unlocked potential of the Lanarkshire coalfield, Ayrshire's coal-bearing hinterland was also relatively unworked. Both counties were, as yet, economically backward, sparsely populated and hampered by poor transport communication, all of which combined to obstruct serious market demand for coal production. Until the deepening of the River Clyde at Glasgow from the later eighteenth century, the growth of trade, of mining and of all other industrial enterprises that required heavy shipping, was severely hampered. The main exceptions to this undeveloped profile by the later seventeenth century were the lower parts of the Clyde Valley coalfield which converged on Glasgow and provided some of its coal supplies; and the coastal sector of Ayrshire with its scatter of salt pans and

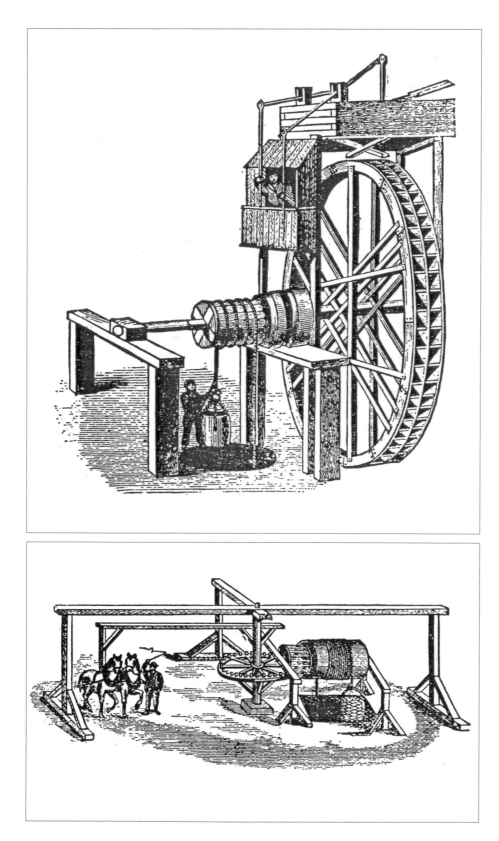

Plate 7.

Water gin, for raising coal in shallow shafts.

Plate 8.

Horse gin, for raising coal and carrying workers.

sporadic coal workings, as on the Eglinton estate, and around Ayr and Irvine. At this time, neither Troon nor Ayr had harbours suited for heavy craft such as collier boats, which obstructed their potential as coal ports.

Although there were no large-scale collieries in Ayrshire before the end of the seventeenth century, the opening up of Stevenston colliery is worth special attention due to the nature of the enterprise and the problems that confronted its owner and workers. Robert Cunninghame, inheritor of the Auchenharvie estate, built salt-pans and developed the Stevenston colliery from the 1670s. He built a serviceable harbour at Saltcoats, intending to expand coal exports to the growing Irish market, particularly to Dublin. A stair pit was successfully sunk in 1678, but difficult geological conditions prevailed here – a characteristic feature of the Ayrshire coalfield – including stone dykes intruding on the coal seams. Moreover, the colliery's low-lying shoreside location was under constant threat from invasion by sea-water and sand. A tunnel one and a half miles long was constructed to drain the higher sections of pit workings on the Auchenharvie lands, but remaining problems of flooding at the lower, shore end limited regular production and prevented deeper mining. Here, as elsewhere, the drive for deeper pits could not be achieved until the fundamental problems of flooding and drainage were tackled effectively. For a significant breakthrough in this direction mine-owners had to await application of the new technology of machine pumping which became available in the eighteenth century.

Establishing Collier Serfdom in Seventeenth-century Scotland

The enterprises of coal mining and salt making in Scotland were intimately connected in yet another way. For nearly two centuries, workers in both vital industries shared the grim experience of exploitation as industrial serfs. Far from being a relic of the lord-and-serf ties of medieval feudal society, collier serfdom, whereby mineworkers became the property of their employer masters, was a new imposition. A series of repressive laws passed by the landowner-dominated Scottish Parliament during the seventeenth century decreed that colliers were bound to their masters and to their place of work.

The reasons for imposing a form of serfdom on the colliers are starkly clear. Owing to rising demand for coal production from the 1580s, more colliery workers had to be recruited into an expanding industry that was widely understood to be extremely dangerous, brutal, dirty, confined and associated with the degraded status of paupers and criminals. From the 1590s the harsh rigour of the Scots poor law included the punishment of able-bodied unemployed men and vagabonds by sentencing them to a lifetime of enforced labour in coal-works. Consequently, labourers in search of paid work who had some choice in selling their labour power were often deterred by the prospect of freely engaging in such debased employment, and stayed out. In the first instance the landowning coal-masters, as members of the ruling class, sought to use state law to secure their existing workforce. By this means, they aimed to avoid a labour shortage brought about by workers leaving for other employment or being poached by other coal-masters who were prepared to offer better earnings and conditions.

The legal binding of colliers and salters was also justified for reasons of national efficiency, concerned with securing ready supplies of coal and salt as vital goods. In his great codification of Scots law, published in 1681, Sir James Dalrymple, Viscount Stair, gave further sanction to the enserfment of colliers and salters. It was, he claimed:

> introduced upon the common interest, these services being so
> necessary for the kingdom, where the fuel of coal is in most parts
> necessary at home and very profitable abroad, and seeing that we
> have no salt of our own, but that which is made by the boiling of salt
> water, salters are also so astricted: so that colliers and salters, while
> they live, must continue in these services.[3]

Viscount Stair did not include lead miners in that vital category of workers. Perhaps this was because their number was so insignificant and, until that time, located at a single centre of production. Only around fifty mineworkers are mentioned at Leadhills by the middle of the seventeenth century, producing up to 300 or 400 tons of ore a year, but this workforce, too, was enserfed from 1607 until 1695. Here, there was no native tradition of expertise in deep lead mining and in the associated craft of smelting. Like coal mining it was known to be a dangerous occupation and had little appeal to newcomers who were needed to swell the workforce. In 1649, Sir James Hope was granted the privilege of binding his workers, and this was enforced until 1695, when Lady Margaret Hope declared freedom to leave employment if workers were not pleased with their conditions. Such a liberal policy may appear surprising, but the explanation is found in the need to keep attracting new workers into the area, and to other locations in Scotland where speculative activities were being conducted into lead mining. The labour problem was solved at Leadhills and at Wanlockhead by attracting English lead miners from Cumberland and Derbyshire to migrate north, and this was successful only by ensuring that working and living conditions were free from an alien and degrading serfdom. As free skilled workers, English labour penetrated into new lead-mining locations and smelting in the eighteenth century, as at the largely open-cast mines at Strontian and Islay and at the Tyndrum underground mine. They entered into 'bargains' with the company employer, consisting of long-term piecework contracts for tunnelling and for tonnage produced.

Industrial serfdom in Scotland was imposed on a minority of the working population. By the end of the seventeenth century there was probably no more than a total of 2,000 mineworkers (including women and children), rising to around 5,000 or 6,000 by the mid-1770s, by which time legal serfdom was being challenged. The terms of the collier bond, as set out in statute law, were unique to Scotland, and merit close attention in their own right. The separate issue of their practical implementation and enforcement will be considered later.

The crucial Act of 1606 contained five main provisions. The first decreed that no colliers could leave their place of employment without written testimonial from their existing master. Any workers abandoning their employment without due leave were deemed to have committed a criminal act as, by absconding in this way, they had stolen their master's property, namely, themselves. If successfully hunted down and reclaimed by the former master within a year and a day, colliers would be returned and liable to corporal punishment, including beating. Any new master who illegally engaged such colliers from another workplace risked a huge fine. The final part of the Act reaffirmed the spirit and letter of the existing poor-law regime by giving fresh powers to coal-masters to apprehend and conscript vagabonds and sturdy beggars into their labour force.

The 1606 Act was extended in 1641 to encompass the binding of all mineworkers, including those who did operational and maintenance work, such as winders (for hoisting up the coal), watermen (for pumping operations) and gatesmen (for clearing the underground passages). Again, the intention was to deter further poaching of skilled and experienced mineworkers, as well as trusted hewers and pit teams.

Also in 1641 and in 1661, harsh new labour laws set out a statutory working week of six full days for mineworkers, with a day of rest on the Sunday, when they were expected to attend worship in church. Bound colliers were also banned from taking customary holidays and only one holiday was declared legal, namely at Christmas. Again, non-compliant behaviour by colliers was subject to fines, corporal punishment and humiliation.

In total, the laws of 1606, 1641 and 1661 were designed to secure and regulate an adequate and scarce labour force. They also sought to obtain the cheapest possible option for the benefit of the coal-masters, reinforcing household and estate custom of exploiting family labour by employing dependent women and children.

None of this legislation explicitly created serfdom by binding colliers for life or even for years at a time, but the intent and effect of the laws, combined with customary practices at the point of making formal agreements between masters and workers, led to enforcement. A collier could sign a bond for a year but be refused a leaving certificate at the year end, and therefore have no choice but to remain or to flee, hoping to escape capture for a year and thereby obtain his freedom from his former master. As a bound worker, the movements of the collier were restricted and controlled by the master, who had the power to remove him and family to another coal-works, and even to sell him on along with the business to an incoming owner. Moreover, once a coal-master had secured his existing workforce, he could then exercise his powers and influence to reproduce the workforce by making the condition of servitude a hereditary one. This was done primarily by the common custom of a collier accepting arles – a present or bounty in money or in kind – as a token of re-engagement. In the case of a married man with a family, the payment of arles bound not only the collier but also his family, effectively tying his children to the coal as the next generation of workers. A gift at baptism formed part of a contract, witnessed by the clergy and before God, under which a man agreed to bring up his child as a collier. Thereafter, although there was no actual legal basis for such an agreement or understanding, bond labour became a child's heritage, which often as not was signified by the appearance of the child's name in the inventory of colliery goods. In such circumstances, a collier and his family could be bound for life or for many years to the same master, and even from generation to generation.

This unrelieved bleak profile of the plight of colliery workers in the long period of industrial serfdom in Scotland was, until recently, the received wisdom. Indeed, some of the first historians referred to colliery serfdom in even stronger terms: slavery was their preferred definition. Some distinctions can be made between the two terms of reference. Unlike plantation slaves, to whom they were at times compared, collier serfs could not be sold off at a public auction; they could hold whatever private property they had, and hand it on; and could not be killed or tortured (although any offending colliers who were beaten or lashed for their sins would have found it difficult to make fine distinctions between this form of punishment and torture). Slaves were not paid labour, while collier serfs contracted for, and received wages and other payments. All the same, under the law, slaves and collier serfs alike had no effective redress against unjust treatment and no legal escape from their bound status. In an influential book, *A History of the Working Classes in Scotland*, the famous socialist politician Tom Johnston repeated the traditionally accepted view of an unremitting degrading servitude suffered by an outcast mining population:

> some of whom wore collars round their necks; they were bought and sold and gifted like cattle; they were wholly unlettered; they lived in colonies ... Their alleged 'privileges' consisted in exemption from taxation and from military service and in the legal obligation which rested with the owner to provide for them in sickness and in old age, and to supply a coffin for their burial.[4]

The first of such claims, concerning the enforced wear of metal collars, remains an emotive and vexed issue, and a test of evidence against repeated assertions. What, for example, are we to make of the only known authentic brass collar, an item in the National Museum (Edinburgh), which was allegedly worn by a collier serf? This collar bore the inscription of the unfortunate wearer: 'Alexander Stuart found guilty of death for theft at Perth, the 5th of December 1701, and gifted by the justiciars as a perpetual servant to Sir John Areskine of Alva'. The collar was fished out of the Firth of Forth many years later, and it is entirely a matter of conjecture how it got there, escape and suicide by its wearer being among the possible circum-

Plate 9.
Gilmerton Colliery, Midlothian, 1786. Detail of plan of coal workings, showing hewer and female bearers with creels: see Plate 14 for complete section. (From reproduction in *The Coalminers*, Scottish Record Office (1983)

stances. It has never been proven whether Erskine's bond serf or slave was actually a collier, or specifically sentenced to mine work, although his master did operate coal mines and, for a while, lead mines from which silver was also extracted. Whatever the state of the evidence in such cases, they are nevertheless a reminder of those laws and practices that conscripted some condemned men, other prisoners and so-called unruly beggars to a life of servitude in and around coal mines. They also reinforce in the public mind the perception of colliery work as a degrading form of labour fit for slaves, serfs, idle vagabonds and convicts alike. In that very year – 1701 – the low legal status of colliers was absolutely confirmed, as they were excluded from the terms of an Act of the Scottish Parliament that sought to prevent arbitrary imprisonment and delays in bringing alleged wrongdoers to court. So, even if it is highly unlikely that born-and-bred colliers ever had to wear collars as a brand of their servitude or as a punishment, the 1701 law appeared to tighten the legal noose around bound colliers, who were pursued and convicted for serious misdemeanours such as absconding from their place of work.

In a series of recent publications based upon new research into a mass of estate papers and colliery records, the leading historian on colliers and salt workers in the period of serfdom in Scotland has made a careful reappraisal of the evidence and the issues. His conclusions have revised many of the assumptions made about the experience of colliery workers in the era of serfdom, and will be discussed in Chapter 2. However, he reaffirms the oppressive nature of the serfdom laws of the seventeenth century and their effects. He acknowledges unambiguous evidence that the practice of life-binding did occur in many parts of the Scottish coalfield, and that workers who absconded were vigorously pursued, and variously imprisoned, fined, humiliated and subjected to corporal punishment. He acknowledges also that the children of coal workers on several estates were considered to be life-bound, thus bringing 'the Scottish experience within the margins of a slave system'.[5]

Early Collier Resistance

There is no denying the grim reality of serfdom, the oppressive legal framework, employer demands and the pressures on mineworkers, especially on those women and children who followed the male colliers into the arduous labour and constant dangers of pit work. Nevertheless, it must not be assumed that all mineworkers were a completely oppressed and entirely powerless section of the labouring classes in seventeenth-century Scotland. In a foretaste of what was to come during the eighteenth century, an earlier generation of colliers were already showing that they were not all passive victims of servitude: docile, submissive and resigned to their fate at the hands of their employers and the force of ruling-class law.

Our knowledge of the history of struggle throughout the world informs us that even slaves and serfs are prone to occasional resistance and revolt, and that laws are broken and ignored, as well as made. It can be demonstrated that this was also the case within the ranks of the numerically small class of Scottish colliers throughout the century before they were released from legal serfdom, partially in 1775 and finally in 1799.

Collier resistance, including early attempts at collective action and strikes, is already evident from the end of the seventeenth century. At the Dysart colliery owned by Henry, Lord St Clair, in 1694, twelve colliers combined to stop working, and resolved to return only on their own terms. Kinneil colliery was the scene of industrial action on at least three occasions,

some lasting for weeks, in 1695, 1700 and 1701. In November 1695, the 'coall hewers whollie deserted their work' for two weeks on the grounds that their annual bounty had not been paid, and which they knew had been granted by other nearby coal-masters. This strike was ended only after the Duchess of Hamilton intervened to order this payment, as the long stoppage had resulted in lost revenue. It is already clear, after close inspection of surviving colliery and estate records, that a century or more before the abolition of serfdom, colliers had learned how to manipulate the market and exploit to their advantage short-term buoyant demand for coal. For example, in 1701, it was reported again that the Duke of Hamilton's colliers were 'something troublesome and mutinous ... especially when ships was waiting upon the coales'. Apparently, colliers used such occasions to negotiate real wage increases, in this instance to win a reduction in the price of meal, which formed a part of their wages. Far from being inarticulate and inert, such colliers knew how and when to act collectively to influence wage levels. The landed coal-masters, struggling to maintain an adequate supply of bonded labour while desiring to expand their mining interests, were to find that such assertive collier behaviour would become more pronounced throughout the eighteenth century.

A Separate Community?

The monolithic picture of the mining community as a caste apart, living in remote isolation, and segregated from the rest of the population, is also questionable for the seventeenth and early eighteenth centuries. While some mining communities were remote, as in Leadhills, the geographical concentration of mining in this period suggests a very different pattern of settlement and residence for the bulk of mineworkers. It points to their location in or near other communities of workers in towns and villages, particularly along both sides of the long coastlines of the Forth and in the nearby landward settlements. For the most part, the mineworkers did not live in isolated company houses or rows, as was largely the case later on in the expanding mining frontier communities that sprang up during the nineteenth century in the wake of railway penetration into upper Lanarkshire, inland Ayrshire and Fife. Instead, as at Dysart, adjacent to Kirkcaldy, at Culross, at Alloa, or at Stevenston on the Ayrshire coast, they tended to reside alongside and intermixed with other workers such as the salters and the other inhabitants, joining them in the ale and whisky shops and taverns, in public worship and in other communal activities. On Prestongrange estate, coal miners lived alongside other workers, such as masons, weavers and tailors. Colliers also mixed with and married into the wider community, although it was customary as well as economic sense for a collier in the east of Scotland to choose a bride from within the mining community to follow him down the pit, to save the expense of employing another woman as his coal bearer.

The weight of available evidence also does not support the notion that bound colliers were denied or deemed unworthy of a Christian burial. Only one example from Fife is given of colliers being treated as outcasts to the extent of being buried in unconsecrated ground. Contrary evidence from burial registers of parishes in the Lothians shows that large numbers of colliers and their families were accorded their final rightful place in a parish cemetery following a Christian funeral.

Protecting Coal and Salt: Before and After 1707

Having pushed a battery of harsh legislation through the Scottish Parliament to regulate their labour force, the landed coal and salt proprietors of the Forth proceeded to gain state protection for their two vital industries. Until the end of the seventeenth century, the Scottish Parliament maintained the monopoly position of the coal and salt owners against the threat of competition from cheaper foreign imports, primarily from the north of England. Then, the prospects of the home-based industries came under renewed threat from Tyneside coal and salt before and during 1707 in the protracted negotiations leading up to the signing of the Act of Union with England. The Scottish economy had been depressed from the middle of the 1690s. Coal exports had collapsed; investment capital had been lost in the Darien scheme fiasco; and dwindling revenue from land had emptied the coffers of the coal and salt gentry, who feared for their prospects.

In this climate of uncertainty, amidst the various contending political and economic motives and manoeuvres that decided the eventual votes on the Act of Union, the Scottish nobility and landed gentry who were salt and coal proprietors took decisive action. Among them were the Earl of Wemyss and the Duke of Hamilton, both with extensive coal and salt-panning interests along the Forth; the first along a stretch of the Fife coast, and the other at Kinneil, Bo'ness.

Starting in 1705, this lobby moved vigorously in the Scottish Parliament to protect their monopoly with a defiant total ban on the import of English salt. The economic interdependence of salt and coal has already been emphasised, and two of the Articles of Union gave landed proprietors the guarantees they wanted on protection for those home-based industries. From 1710 until 1793, the coal trade of the Lothian and Fife colliery owners was

granted protection by the imposition of tariff duties on outside competitors. The protective salt duties were also reaffirmed and stayed on the statute book until 1823. Both measures undoubtedly helped coal producers along the Forth to remain in business throughout the eighteenth century, to employ a stable, although unfree, workforce, and to service home demand in the east. However, before the end of the century their long period of prominence within the mining profile of Scotland was to be successfully challenged by the rise of the western coalfield region and new sources of demand for coal during the first phase of the industrial revolution.

Masters and Workers in Mining: from Landed Estate to Industrial Revolution c.1707–1830

Coal Mining: Demand and Expansion

The rapid progress that was made in every department of the coal business during the eighteenth century is quite astonishing, and the plan of operations has been reduced to a regular system. Bold and enterprising schemes have been put in execution; and we now find in Scotland pits of 140 yards, and in England no less than 300 yards deep. The great and prominent improvements are: The introduction of the steam-engine for drawing water, the improvements of which are such, that it is capable of doing work to any extent, and is applicable in all situations.

The introduction of the rotary steam-engine, for drawing coals up the pit, invented about the year 1782. The saving of expense which has been produced by its means is very great; for, without it, the expense of drawing coals with horses from the depth of 300 yards, would have tended very much to raise the price of coals.

The execution of work performed by this machine far outstrips all former ideas we had of drawing coals quickly, as it is no uncommon thing now to see a basket of coals weighing 7 cwt. drawn 200 yards up a pit in the space of fifty seconds.

The introduction of waggonways for transporting of coal from the pit mouth to the shipping place, particularly the recently improved cast-iron rail-way: formerly it was the work of a horse to bring down 6 cwts at a time; one horse will now bring down from twenty to thirty times that weight.

The last and greatest improvement in point of humanity is the substituting horses, in place of women, for transporting the coals below ground; and there is ample room for great exertions in this department, both on the River Forth, and in Mid-Lothian, particularly in the latter district, where very few horses have been introduced.[1]

The coal-mining industry in Scotland underwent rapid expansion between

1700 and 1830, recording a significant and continuing rise in output from the 1760s. Although historians dispute the size of total outputs for this period it would appear that, from a realistic low base of around 225,000 tons a year in 1700, coal production leapt to around 2 million tons by 1800, and to over 2.5 million by 1830. Overall, this growth represented a tenfold increase in production over the entire period, and 13 per cent of the share of total production in the British coalfields.

Within this period, which spanned the first phase of the industrial revolution from the 1780s, the principal direction of Scottish trading activity switched dramatically from its historic eastern seaboard connections with English coast markets, northern Europe and the Low Countries. Following the Treaty of Union, and access to Atlantic and American raw materials and markets, new outlets for trade opportunities opened up for entrepreneurs in the west, in particular for the Glasgow area and the Clyde ports. By the end of the century, much of the fortunes made from overseas trade in slaves, sugar, tobacco and cotton was being invested in coal-bearing landed estate, in the cotton industry and in the iron manufacturing industry.

The first coal-smelting iron works at Carron, near Falkirk, in 1760 signalled the arrival of heavy industry, but not the immediate onset of the industrial revolution in that sector. Carron remained the only large iron works for twenty years until others were established, mainly at inland locations in Lanarkshire and Ayrshire between 1779 and 1801. They included Wilsontown, in remote upper Lanarkshire, in 1779; Clyde Ironworks, near Glasgow, in 1786; Omoa (Cleland) and Muirkirk, both 1789; and Glenbuck, Calder and Shotts between 1795 and 1801. Devon, near Alloa, and Balgonie, Fife, completed a total of ten iron works by the start of the nineteenth century. Their combined blast furnaces produced 22,800 tons of pig iron in 1800, 9 per cent of the British total, rising in later years to an annual total of around 30,000 tons. Stimulated by high prices in 1825, the Chapelhall and Monklands (Calderbank) iron works were built, followed in 1828 by Gartsherrie, at Coatbridge, where the Bairds began to lay down the first industrial plant of their great iron empire.

Plate 10.
Horse-drawn coal railway, eighteenth century.

All the new iron works before 1830 required large quantities of coking coal for the smelting process and were preferably sited at or near mineral-rich locations where ready supplies of clayband ironstone and coal could be mined together. Each ton of pig iron required ten times that amount of coal consumption: for example, Muirkirk in 1830 was using up to 36,000 tons of

coal, supplied mostly from its nearby Kames colliery. However, although the
fuel needs of iron manufacture stimulated the opening and development of
many collieries, the cold-blast method of heating, combined with poor-
quality coking coal and local iron ores, produced inferior and costly pig iron.
This largely explains why market demand did not grow greatly until after the
revolutionary hot blast method was introduced into the furnaces from the
1830s. Applied to the quality blackband ironstone and splint coal, both of
which were available locally in great abundance, the hot blast brought a new
vitality to an industry which was soon to create in the Monklands area of
north Lanarkshire a world-famous, first-class pig iron. The dramatic rise and
development of the closely-connected coal and iron communities of this part
of the Lanarkshire industrial frontier in the 1830s and 1840s is appropriately
dealt with in another chapter.

By the early decades of the nineteenth century, application of the
invention of steam-powered engines revolutionised production in the textile
industries, particularly in the spinning process, No longer dependent on
water-power locations for spinning and weaving, by 1830, over 100 steam-
powered cotton mills proliferated, principally in and around Glasgow and
Paisley. The fuel needs of Scotland's largest thriving industry gave an
enormous boost to coal demand and stimulated further development of
mining, especially in lower Lanarkshire (which then included Glasgow) and
into nearby Renfrewshire. As will be seen, within the Scottish coal industry
(although less so in the eastern coalfields), the introduction of expensive
steam-powered engines for drainage and winding was to make a significant
impact on capacity, output and working methods within the deeper mines.

Plate 11.

Gartsherrie iron and coal
works, and part of
Glasgow and Garnkirk
railway *c.*1832.
(*North Lanarkshire
Council*)

Throughout the eighteenth century, mainly along the Forth, colliery proprietors had laid down wooden and metal waggonways, sometimes stretching for several miles, to take coal down to ports and harbours. This coal transport was horse-drawn, but by 1830, in some locations, this mode of power was beginning to be replaced by moving engines on iron rails. Here, at the very dawn of the railway age, the early potential of steam-powered railway locomotion for haulage was being piloted at the colliery pithead and for outward transport of coal on short-distance mineral lines. It was no co-incidence that the birthplace of Scotland's modern railway system was the mineral-rich Monklands. The first line authorised to use a steam locomotive for haulage was the Monkland and Kirkintilloch railway, opened in 1826, followed by the Ballochney Railway, both lines being designed for coal transport into Glasgow, the principal coal destination in Scotland.

In accordance with this re-orientation of capital investment and the emerging coal-fired needs of industry by the early nineteenth century, the centre of gravity in coal production was shifting from the eastern to the western coalfields. The old-established active coalfields of Fife and the Lothians continued to produce huge outputs of quality house coal for customers at home and abroad, and smaller coal for salt pans and lime kilns. However, it was clearly the phenomenal increase in population and urban growth, particularly in and around Glasgow, Paisley and the lower Clyde valley; and the opening up of new economic developments in the counties of central and western Scotland which set the pace for local output of household and industrial coal. With a population of 80,000 in 1801, rising to 200,000 by 1830 and still growing apace, Glasgow had become the commercial and industrial centre of Scotland. Glasgow merchants and indus-trialists financed the 12-mile-long Monkland canal, completed in 1793, which was intended to pour cheap coal into Glasgow along the canal route from the developing coalfield of the Coatbridge and Airdrie districts. However, the real worth of the canal was not fully exploited until the 1820s, when short cuts were made to the ironworks at Calder, and the Gartsherrie, Langloan and Dundyvan works at Coatbridge. Only then did a constant volume of coal and iron begin to move out to the Clyde and to distant markets.

In this transformation, Lanarkshire became the primary expanding coalfield, while the Ayrshire coalfield also registered a rising profile, concen-trating on coal exports to Ireland. Increasing production of coal in the Stirlingshire and Clackmannan coalfields showed a continuation of existing sources of demand and, to some extent, reflected greater industrial demand for coal, as in the fuel needs of the iron industry at Carron and Devon.

Estimated annual coal production by 1800 across the Scottish coalfield demonstrates the quantity and distribution of this expanding output:

Estimated Annual Coal Production by 1800 Across the Scottish Coalfield[2]

Region	Production (in tons)
Lanarkshire (with Renfrew and Dumbarton)	550,000
Lothians	500,000
Fife	350,000
Ayrshire	250,000
Stirlingshire	200,000
Clackmannan	100,000

Mining Methods and Hazards

Few technical improvements in efficient mining methods were introduced during this period, although some, notably in steam-powered drainage and winding, were important, and had their effect on work processes and working conditions. In eighteenth-century Scotland, the dominant

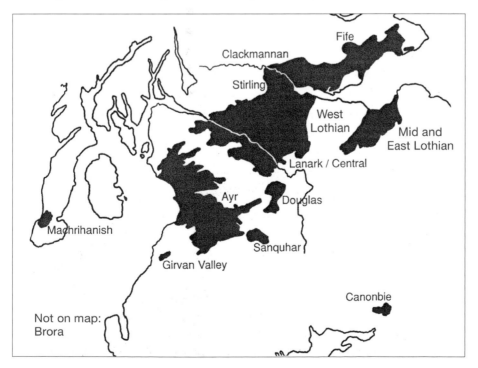

Plate 12.

The Scottish coalfields.

landowning coalmasters invested in improved technology for draining their collieries. Later in the century, when coal companies came on to the scene, they, too, were prepared to do likewise. The big landowners and the new companies were usually better placed than the lesser coalmaster gentry to afford the considerable expense of installing and maintaining the Newcomen steam engines. An engine-house, chimney, pumps and labour were added to the costs of a steam engine, and if coal costs are excluded, installation expenditure alone was above £1,500, before operational expenses. The Clerks of Penicuik, for example, who maintained a profitable colliery at Loanhead, calculated against buying one or more Newcomen engines. They decided against using this method to sink deeper pits, reckoning that the viability of their small-scale enterprise would be endangered by the cost of the outlay on this new machinery and on its upkeep. Instead, they chose to rely on horse gins. On the other hand, among the large landed estate mining enterprises, the Stevenston collieries in Ayrshire, and Hamilton collieries at Bo'ness were well equipped with steam engines and thus were more productive in sinking deeper pits. At Bo'ness, in the 1770s, Newcomen engines enabled effective pumping to be installed in the deepest pit in the country (70 fathoms, or 420 feet).

The number and rate of Newcomen pumping engines introduced into the

Plate 13.

Remains of a Newcomen atmospheric steam engine.

Scottish coalfields throughout the eighteenth century should not be exaggerated. The cost factor was obviously vital, but deep mining was impossible without the new technology. The first Newcomen engines were applied to mines in Scotland in 1719, and yet only six were installed by as late as 1759. Twenty more were introduced in the 1760s, indicating greater investment in deep mining. This capital commitment is even more evident from the 1770s onwards, and by 1800 around eighty steam engines were pumping at the larger mines, primarily in the west of Scotland.

In many collieries, horse and water gins for pumping were still the order of the day – the best of the endless chain and buckets system managing to drain to around 30 fathoms – but such means of drainage could also be expensive undertakings. The same was true of the many drainage levels driven into hillsides in all the coalfield areas. But neither gins nor levels could drain the deeper pits like the Newcomen engine and its more efficient, less expensive successor – the rotary steam engine designed by James Watt, which was introduced into the Scottish coalfields only after 1800.

The introduction of steam winding in Scottish collieries belongs to the 1790s but, by the 1830s, was still largely confined to the deeper mines. Moreover, steam-powered winding was not in use in those stair and gallery pits where the edge coal seams were on a steep incline and sometimes nearly perpendicular, as in many Midlothian and East Lothian pits, which remained technically backward. Elsewhere, in the absence of this improved method of winding, in some collieries, as at Fordell in Fife in the 1790s, this process still relied on horse gins, with female bearers taking the hewn coal to the hoist position at the pit bottom. Rather than steam-powered machines, horses were in common use for both pulley winding at the pithead and for underground haulage purposes in the main roadways by the end of the century. Thus, for such tasks, where horses and ponies were being introduced as a general substitute for manual labour, there was less need for women and child bearers underground. As will be seen later, women bearers were virtually unknown in the more progressive mines of the west, a fact which had much to do with willingness to invest there in the greater efficiency and productivity of horse and machine power. Where difficult geological conditions prevailed, as in the working of the sharply inclined edge seams in the east, the owners of those collieries argued that the manoeuvring of horses to pull hutches or wagons along underground roadways was not feasible in such sloping conditions; and they generally continued to fall back on the exploitation of human labour for the unforgiving tasks of bearing and haulage. There were growing numbers of large and progressive colliery complexes by 1830, but the more prevalent unit in Scotland was the small,

Plate 14.
Overleaf. Gilmerton Colliery, 1786. Plan by John Ainslie. The schematic section shows hewers at coal face; bearers carrying loads to pit bottom along roads and stairs; and bottomers raising coal to surface by rope and bucket.
(The Coalminers, *Scottish Record Office* 1983)

ECTION *of One of the Seams* N.B *the Stoups & Coal are Shaded B*

backward colliery where a single horse gin for either pumping out water or taking up coal was the only investment in machinery.

In addition to the acute problems of drainage, as underground workings extended, there was an increasing risk of explosions from accumulations of firedamp (methane gas) at the working faces. The loss of life and property incurred by such rare, but violent explosions, is shown from this incident in 1805:

> A dreadful occurrence took place on Monday morning at Hurlet coal work near Paisley. About 9 o'clock, while the men were at work, the inflammable air in the pit took fire. Four men were blown from the bottom of the pit into the air, the mangled parts were scattered about in all directions. One of them was found at the distance of 300 yards from the mouth of the pit. There is every reason to believe that the other thirteen, who were below, have all been killed. The father of one of the sufferers went down in the hope of saving them, but was instantly suffocated by the foul air. A horse at the mouth of the pit was killed and the whole of the machinery blown to atoms.[3]

Exposure to blackdamp caused severe breathing difficulties and suffocation, and its presence was tested by the primitive method of noting the effects of this gas on a small animal such as a bird or on a naked light. If both were seen to struggle and die quickly, this was the signal of immediate danger. To combat such gases, various means were adopted during this period to improve the circulation of air through the underground roads. These included air shafts; and upcast and downcast shafts in which the flow of air was generated by bellows and furnaces at the pit bottom. Child trappers were also employed to open and close ventilation doors inserted into the roads. However, the installation of furnaces and additional air shafts was not very evident in Scottish mines during the early decades of the nineteenth century, with the exception of a few large collieries. On this aspect of safety, comparing the provision of costly ventilation equipment in the large Tyne and Wear coalfield where poisonous and inflammable gases were more prevalent, the leading historian of the Scottish coal industry for this period passes a scathing verdict on the callous standpoint of the owners:

> In Scotland, the occasional death of a pitman from blackdamp was not seen as reason enough to invest in expensive ventilation.[4]

In the same area as the explosion incident described above, one of the first recorded instances of pioneer installation of mechanised fan ventilation was in 1827 at a colliery near Paisley, where the air pressure was provided by a steam engine. This, at least, marked the beginning of modern technology in mine ventilation, although it is worth repeating that the many small mines in Scotland well into the nineteenth century operated no ventilation aids at all, condemning thousands of workers to terrible illnesses and premature death. To what extent any of the newly devised safety lamps were in use in Scottish collieries in the early nineteenth century is not known. An interesting description of the introduction of the Davy safety lamp, fitted with wire gauze to shield the naked flame from encounters with mine gas, comes from inspection of a gassy mine at Alloa in 1816. This account gave the new safety lamp a vote of confidence:

On reaching the pit bottom, we set out for the forehead with common candles, and proceeded fully 300 years ... till we got to the end of the air course. At this point I was told that we could proceed no further with common candles on account of the air being very inflammable. At the firing point we therefore took safety lamps and proceeded to the forehead as we went forward, the effect of the lamp was clear and distinct, the flame was greatly elongated, and reached to the top of the lamp. The strength of the flame was evidently increased as the lamp was held near the roof, and it gradually shortened when the lamp was placed on the pavement, and allowed to stand there for a short time. On arriving at the forehead of the mine, I found several colliers at work, pushing forward this mine to another pit, each of whom had one of these lamps close to him which yielded abundant light for his work. I said to one "Well, friend, is not this invention a great blessing of providence granted you to prevent our lives being suddenly taken away?" – he said "It certainly was so ... and could not think enough of it". Another said "Sir it is worth twenty lives". I asked another if he had any objections to go to the most inflammable and dangerous part of the mine with the lamp. He answered he would go without the least hesitation to any part. He knew that the lamp would keep him perfectly safe provided the air was fit for breathing ... I requested that we might go and seek such places as the inflammable air would be most abundant, this we did with full confidence and safety – and I now state with pleasure and satisfaction that I would have no hesitation to descend into any mine however inflammable with Sir Humphrey Davy's Safety lamp. I know my life would be in safety as to an explosion.[5]

Plate 15.

Testing for gas with the Glennie lamp, a later version of the safety lamp. (*North Lanarkshire Council*)

There was, as yet, no official inspection of mines in Scotland, and no government regulations placed on employers to record and account for mining accidents and deaths. Nor was there any statutory requirement to provide basic safety equipment or protective devices. To employ colliery labour was cheap; but the price of coal in human life was already extremely high, and would continue to be so.

In the case of lead mining, much the same technological advances as in coal mining were evident at Leadhills and Wanlockhead. By 1800, rapid expansion of production in this area reached around 5,000 tons of ore in the best years, most of it going towards the manufacture of paint and piping. Technical improvements in drainage mirrored those in coal mining, making it possible to work at greater depth: the deepest lead mine in 1800 was 120 fathoms. A water-powered pump engine was in use at Wanlockhead by the 1720s, and one of the earliest Watt steam engines to be brought to Scotland was installed here by the 1780s. Coal supplies for the mine engines and the smelting furnaces were transported from the nearby Sanquhar mines. Three steam engines were operating at Leadhills by 1782, although this expensive method of pumping was abandoned during years of depressed production in the late 1820s and early 1830s.

Here, the labour force underground had never included female workers, and pit boys started work at a later age than in coal mining, working with adult males in dangerous conditions and in a highly polluted atmosphere. Apparently the lead miners did not have to contend to any extent with the dangers of firedamp or blackdamp. However, into the nineteenth century, frequent accidents, combined with lead poisoning and lung disease from constant inhalation of rock particles from regular blasting operations, produced at least the same level of fatalities and premature death as coal mining.

Finally, in this brief review of the extent of technological innovation in mining during this important formative period, it is worth re-stating that the essential primary task of cutting and extracting the coal, the ironstone, or the lead ore, was still accomplished by the manual labour of the hewer. Shot-firing – using gunpowder for initial blasting – had begun by 1830, but it was not yet generally practised, except in the lead mines. The pitman wielding hand tools was the principal production worker underground, whether it was the single collier in a small, primitive pit; or, as in most pits, the hewers working alone or with an assistant hewer at the immediate coal face within the stoop and room arrangement. It was also the case from the 1760s in some of the larger collieries, where the small team of hewers worked in a co-ordinated fashion side by side in a stepped line along the length of a coal face. However, in noting this alternative to stoop and room working, we

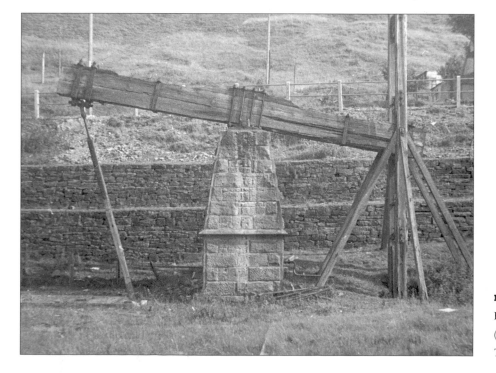

Plate 16.
Beam engine, Wanlockhead.
(*Wanlockhead Museum Trust*)

need to distinguish this emerging version of longwall working from the method of longwall as practised and understood in the twentieth century. In the modern era, the machine cut coal in wide strips, advancing ruthlessly in a single line through the seam.

Masters and Mineworkers: Work Discipline and Labour Relations in Eighteenth-century Scotland

I loosed the whole coaliers and bearers of Lonhead from their work upon Wednesday the 20 of January 1703 till they should condescend to the whole following articles.

1. to guard against all profanness and immorality particularly against excessive drinking and tipling, fighting and flyting, cursing and swearing and taking the lords name in vain and in case of failzie to make their publick acknoulegement and Confession before the Congregation of Lasswade and to submit themselves to the censure of the Kirk Session as they shall be appointed by the Minister and Elders.

And to be liable in payment of their fines imposed or to be

Plate 17.

Lead mines at Wanlockhead. J Clerk of Eldin: contemporary drawing, *c.*1775. (*Wanlockhead Museum Trust*)

imposed by the Master conform to the acts made by him against all such profanness and gross abominations.

2. The hail coaliers shall work whole days to witt twelve hours space each day of the week at least so many hours dayly as may furnish to their bearers twelve hours work dayly to bear out your masters wrought coals.

3. That every coalier keep and employ as many coal bearers as will be suitable to the coal-wall and to the coalier's strength.

4. That every coalier pay dayly 5/- Scots to each bearer who are not hired for meat and fee within there houses.

5. That every coalier pay to the master (for each day he lies idle) as much money as by his work he hath gained to him upon any other work day of that or the preceding week Except the coalier get liberty to ly idle of yt, day either from the master or in his absence from the Oursman and Grieve both together or that he be really sick as an sure evidence … Because for the most part it hath been found that such as lay idle did nothing but go up and down seeking people to drink and tiple with them.

6. … the coaliers do oblige themselves … to lay quite aside the unprecedented and unusual way of working quarter and half days which hath creeped in amongst you to the loss of all parties and which is inconsistent with the old Custom of your place.[6]

The document above, signed by Sir John Clerk, landowner and master of Loanhead colliery, seventeen colliers and two witnesses, indicates the nature and some of the problem aspects of the employer–worker relationship in an estate colliery under serfdom. It is by no means a contractual agreement between masters and workers; rather a decree drawn up by the master, demanding obedience from the workforce. It sets out details of the imposed work schedules, including a full six-day working week and the required accompanying discipline. Regulations 2,3 and 4 are of special interest in that they specify the conditions whereby the male hewer has either to supply female bearers from his own family or hire another woman or girl to carry the coal out of the pit. The document also stipulates a code of conduct intended to instil and enforce moral discipline and good, orderly behaviour among the colliers and the women and child pit workers. It also carries strong overtones of a strict Presbyterian regime, emphasising the virtues of hard work and godliness, with personal conduct answerable to the rule and powers of the landowning coal master and his allies in the minister and elders of the Kirk Session in Lasswade parish. The letter and tone of the dictated agreement conform to statute laws of the seventeenth-century governing serfdom (especially the ruling on a six-day working week, which Clerk's colliers had apparently been evading) and the force of local, estate, customary arrangements relating to work and conduct.

Sir John Clerk I (died 1722) and his son, Sir John Clerk II (1676–1755) were among the few members of the landed gentry who undertook direct management of their estate colliery. Typically, in such cases within collieries, the owners delegated duties to their operational managers. The document above mentions a grieve, who was the general on-site foreman, and the oversman, who was in charge of the underground workers and operations. As owners and coalmasters, the Clerks were also rare in their genuine and knowledgeable interest in mining affairs, and Clerk II, besides being a leading politician and trained lawyer, had a deserved reputation as an expert in techniques, methods and economic aspects of mining. However, like his father and his son James, he had a trying time as an employer of colliery labour. It was not only the colliers who continued to prove troublesome, as the Clerks had to contend with several grieves and oversmen who were apparently incompetent and dishonest, indicating a general problem of the initial lack of experienced and capable managers in the eighteenth-century Scottish coalfields.

Landed estate coalmasters like the Clerks, and the coalowner nobility above them, had several sources of formal power at their disposal to discipline the workforce. They could resort to statute law of the previous century in attempts to enforce the terms of the serf bond; bring labour and contract disputes with other employers before the Court of Session in Edinburgh; apply for immediate assistance from the sheriff to pursue and bring to trial any deserters; as feudal overlords, use their own barony courts to convict and punish offenders; use the disciplinary powers of the Kirk Session; and, as employers, exert their undoubted authority to hire, fire, physically punish and banish any troublemakers. Sir John Clerk II was known to have used this whole array of powers and sanctions against his workforce. It is possible to judge him as a strict but fair master who applied a 'carrot and stick' approach to labour relations, and not as an unduly harsh employer by the standards of the time.

From the very full record available for the Clerk regime in the eighteenth century, it would appear that, as estate coalmasters, they rewarded positive attitudes and regular hard work. Notably, they did not pay the lowest wages, perhaps with an eye to their own self-interest in retaining a motivated workforce.

Some miners who could not face such imposed conditions and

suffocating moralising controls on their life opted to abscond, and took the risk of recapture and punishment. In 1702, John Kirkwood, a young bound miner at Loanhead, failed in his attempt at escape, and in typical Clerk style, was obliged to sign a new bond incorporating the following pledges:

> To study and know the principles of Christianity more than ever I did formerly, to learn my catechism, to learn to read and write, and to be more frequent in prayer than formerly ...
>
> To be at my work daily by four in the morning.
>
> To refuse to do nothing which either the said Sir John or his Lady shall command.
>
> To go to bed every night by eight a clock at furthest.
>
> Never to leave the said Sir John Clerk his work without his consent and till he provide no work in some other place.[7]

The stern paternalistic stance of the Clerks was fused with a concern for the basic wellbeing of their workforce, in which sense they were prepared to recognise their feudal obligations to attend to their workers and dependants who were ill or frail. Houses were provided at Loanhead at a low rent, with a supply of piped water to a stand pump, and free weekly allowances of coal. In character with this employer regime, perhaps more so than in some other estate collieries, sobriety, thrift and self help were encouraged among the colliers, including a sickness and funeral benefits society, with subscribing members.

Although labour discipline under serfdom was often severe, some colliers had scope to evade the letter of the law. For example, on the issue of collier absenteeism from work, as indicated in the wording of the document above, the working week of six full days laid down in the statutes of 1641 and 1661 could not be uniformly enforced. Of course, there were always solid practical circumstances which happened to interrupt the rhythms and patterns of normal working, and prevent anything like the implementation of a regular six-day week. Enforced stoppages of this nature included physical and environmental problems such as inadequate ventilation, unexpected rises in water levels and encounters with rock among the working seams, while human aspects such as injuries and illness would also intrude. However, other reasons for stoppages and shorter working arose from live tensions and conflicts between masters and colliers. For instance, at Bo'ness, a set of new regulations adopted in 1740 was aimed at forcing the colliers into working 'all the six days of the week, and not lye idle one day of the week, as they have always done hitherto'. Masters appeared to be more concerned about obtaining long periods of regular work from their colliers, as in settling the length of the working week, rather than the length of any working day. In 1769, John Burrell, the Duke of Hamilton's leading steward, inspected the colliery at Bo'ness, where Doctor John Roebuck was the tenant and manager. He found that the colliers were successfully resisting the regulations which had been made in 1740:

> None of them would submit to the Doctor's fines for not working five days of the week. I at length told them that as the law required them to work the whole six days of the week under the pain of 20d, I thought it no hardship that the Doctor was putting upon them to work five days of the week under the pain of 1 shilling, and further advised them to make tryal of it for two or three weeks … and if they behaved as they ought to do should find me their real friend and if they behaved on the contrary they might depend on me being their moderate enemy …[8]

Despite the cajoling and veiled threats coming from one of the most powerful men in the country, at Bo'ness as at Loanhead, and from what is known about other estate collieries, it would seem that five days was a typical colliers' working week for most of the eighteenth century.

A working day of at least ten hours is found quite widely across the Scottish coalfields for this period, although some colliery regulations specified twelve hours, and there were undoubtedly estate mines where an even longer day was in force. However, as at Loanhead, colliers were not necessarily tied down to a ten- or twelve-hour working day. In practice, at most, it clearly meant cutting sufficient amounts of coal to keep the bearers employed over a twelve-hour shift and it was the bearers who had to endure the longer working day. What was most at issue between the masters and the hewer was the total reward for the hewer's work. It has to be appreciated that even life-bond hewers were nevertheless assertive and skilled workers who could often negotiate part of their terms of employment, especially wages. As by custom, hewers were not paid by the day, but by piece rates, depending on their output, and they in turn were responsible for paying or providing the upkeep of the team of bearers. Moreover, good, experienced hewers were in short supply. Their availability had a scarcity value for a master and many hewers found that they could exert this position to their advantage as a bargaining counter over wage rates and earnings.

Those bound colliers who, for whatever reasons, did not, or could not, find such scope to improve their conditions, had no real choice other than to endure imposed conditions. Where masters were able to dictate terms, fix piece rates at a low rate and insist upon a quota of output, this meant that hewers had to work more intensively or for longer hours to achieve an acceptable, reasonable return in wages. Such miserable oppressive conditions did undoubtedly prevail, and especially in more isolated, inland pits where demand for coal was seasonal and poor, and where consequently, the bargaining power of the collier was very limited. These circumstances existed at collieries such as Carnwath in Lanarkshire, where work was uncertain throughout the year. This could result in extra worry and deprivation for bound workers if their legal owner was unwilling to lend them to a neighbouring master. Moreover, if alternative work opportunities were not available, colliers and their families would have to resort to begging, charity, or dependence on the poor-law system.

However, a considerable body of evidence suggests that, individually and collectively, bound colliers in eighteenth-century Scotland had definitely begun to voice and develop their own notions about what constituted a day's work, what they should receive as payment and the amount and use of time outside their working hours.

It is not surprising to find that bound colliers had a preference for a shorter working day, and that some were prepared either to force the issue with their masters, or alternatively to adopt passive resistance and hold out until the master came to acceptable terms. Such collective intent and action could well result in effective limitation of the day's work and time, simply by choosing to stay off work. To the collier, sufficient rest, a reasonable and flexible length of working day, time off and some leisure were due rewards for his stints of hard physical labour. To the master, short-time working was perceived as a challenge to his authority; an interference with production and profits; wanton idleness and an excuse for indulgence in drink or other misdemeanours. The independent-minded collier aimed at limiting the day's work according to his own notion of adequate daily output, and tended to resist any expectation or drive to make bigger earnings and output where this extended effort meant longer working hours. This concept of an adequate day's work, usually expressed and understood as a quantity of output, arose during the period of serfdom, and is known in Scotland as 'the darg'. Needless to say, conflict over the changing meaning and value of the darg involving collier practices in restricting output would be an abiding ingrained theme throughout the long and troubled experience of labour relations within mining in Scotland

Even under the bond system, the collier's assertion of a balance of work and leisure was aimed at gaining some freedom to participate in ordinary life.

And although attempts to paint a more optimistic picture of working lives in coal mining communities may not represent the experience of the majority of colliers, the archive record provides some examples of a more positive outlook. Snapshots of the experience of mineworkers at Urquhart colliery in Fife in the 1740s may be taken as one example. Here, the six-day week was generally ignored and colliers often took days off. Monday was the most likely day for volunteering to stay off work, while Saturday was usually the best-attended working day, as colliers strove to maximise earnings before their weekend. There is also evidence that they tended to take days off when their piece-work earnings were higher than normal. In those circumstances, colliers would choose to work a three- or four-day week, giving rise to the absenteeism commonly encountered by masters elsewhere. At Urquhart, attendance and output also suffered due to the participation of colliers and their families in the annual cycle of holidays in lowland Scotland. To the fury of their masters, virtually all work stopped for two or three days at New Year, as the workers went on a round of visiting neighbours and friends. Yule and Handsel Monday were also widely celebrated. In 1744, a group of colliers feigned illness to go to the Dunfermline Fair, and were away for two days. Baptisms, weddings and funerals were also occasions for leaving off work for up to two days at a time, often with opportunities for convivial hard drinking. Other touching aspects of family duty and affection figured in unpaid absence from work, as in the following examples. William Bevrage was absent for four days, tending to his dying child; David Bevrage stayed off work to help at home for three days while his wife was giving birth; and Robert Wilson stayed out of the pit for four days, to make arrangements after the death of his brother's wife, and find a nurse for the distressed children. In this bonded community, at least some semblance of a normal life, family obligation and a temporary welcome relief from the hardships of work could be observed.[9]

Other evidence which counters the image of totally downtrodden, docile and cowed collier serfs in the eighteenth-century coalfields is to be found in the many known instances of collective bargaining with employers; in direct action, including strikes; and withdrawal of labour by absconding from the workplace. Frequent references to desertions from pits, especially when they involved concerted action by a group of colliers, were little or no different from going on strike. Outright strike disputes during the first half of the century occurred at the Duke of Buccleuch's Cowden colliery, Midlothian, in 1728; at Rothes and at Urquhart, Fife, in 1742 and 1744; at Elphinstone, East Lothian, in 1743; Kilwinning, in 1747; and Irvine, in 1749. Such localised workplace attempts to influence or improve wages and conditions often required courage and determination against formidable odds and, far from being desperate, despairing, spur of the moment acts, they showed a calculated awareness of when and where to register a protest and to threaten or take industrial action.

For instance, in 1747, at one of the Earl of Eglinton's pits near Irvine, the colliers had combined to win a higher rate for coal cutting. This, in turn, prompted neighbouring colliers to demand the same rates, supposedly 'being disposed to mutiny insisting that their Wages ought to be raised to the same standard with that of Eglintoune'. When in 1749 the Earl of Eglinton's new grieve tried to reduce the locally recognised cutting rate he was met by 'general Combination to abandon the works at once'. This direct action ended in defeat and the reversal of earlier gains made by the workers. In such cases at busy collieries, sooner or later, colliers would renew the struggle for improved conditions.

Militants could expect no mercy when open confrontations with masters ended in defeat. Lord Elphinstone ordered ringleaders of the 1743 strike to be banished from the area. However, as if to indicate the turmoil in the late eighteenth-century labour market, Robert Reid Cunninghame, owner of one of the most technologically advanced collieries in Scotland at Stevenston in Ayrshire, had banished 'forever' several men in the early 1770s, on the grounds that they had been 'at the bottom of all combinations'. Yet, he was obliged to re-employ them four years later, such was his desperation to overcome the labour shortage of experienced colliers.

From its inception, the Carron Iron Company adopted a hard-line policy against recalcitrant and striking colliers, fearful of severe disruptions to industrial production brought about by labour unrest. In one incident during the 1760s, troops were called in to quell a disturbance by angry colliers. In 1775, when there were ' Combinations and Tumults' at the collieries, troops were again summoned, workers were disciplined, jailed, dismissed and evicted from the company houses. As will be seen in later chapters, this was a pattern of action and reaction which would often be repeated in the history of major confrontations in the Scottish coalfields.

The frequent occurrence of labour troubles at Carron, where over one hundred colliers were employed at nearby pits, shows that employers could not easily prevent collective organisation among their colliers. Even in scattered mining communities, it can not be readily assumed that the doings of colliers, both at the workplace and outside it, were strictly supervised and controlled by masters and grieves. When considering the existence of worker organisations, it has to be remembered that, during this period and until the

mid-1820s, colliers who met for the purpose of taking decisions and collective action on employment and work issues were committing an illegal act. Combining or forming combinations to promote such activities were illegal under Scots common law, even before the Anti-Combination Laws of 1799 were legislated in the British Parliament. The same ban applied to any workers' meetings or organisations formed for political purposes, especially of a radical nature. During the 1790s, the influence of radical ideas stemming from the French Revolution and home-based ideas and organisations which promoted democratic rights, prompted the government and propertied interests into harsh suppression of such movements among the working classes. Consequently, whether deliberating on work grievances or political issues, colliers were likely to meet in secret, sometimes underground, away from prying managers and foremen, or under the guise of a friendly society or sickness benefits club which was recognised as a legal activity. At such meetings they made their own bonds of solidarity, taking and keeping solemn oaths of craft brotherhood. This semi-masonic 'brothering', which was to become increasingly influential among colliers towards the close of the century, was not unusual among skilled workers elsewhere in Britain during the eighteenth century. Having no political or negotiating rights, oaths and grips were a practical means of recognising fellow workers and providing mutual support against masters and other hostile forces.

Whether at the level of informal or formal organisation, colliers discussed and acted on labour issues. Apart from those issues already mentioned above – exerting control over wages, hours, the pace of work and output – another major concern was to protect against entry of unwanted 'strangers' into pit work. In particular, they sought to guard against incoming labour which threatened to compromise or dilute agreed work practices and conditions. Separate evidence from Midlothian in the 1720s and from Ayrshire in the 1770s points to the earliest existence of collier brotherhoods committed to restriction of entry into the pits. Again, this policy of defending craft interests – including the eighteenth-century equivalent of the 'closed shop' and 'working to rule' – posed a major challenge to the authority of the employer, and was to be a longstanding source of conflict in the mines.

As the mining industry expanded, the supply of labour increasingly became a crucial issue for the coalmasters and was to reach breaking point from the 1760s onwards. Until then, shortages of hewers and bearers were usually local and temporary. Short bond colliers and female bearers alike could be attracted from other pits, sometimes with bounties and higher wage offers. The prospect of continuing in gainful employment was an issue for those mineworkers whose normal workplace had ceased production. Moving on became possible where masters were willing to transfer workers (although not release them from their bond), on account of such reasons as the working out of accessible coal, other technical problems, or periods of slack demand, in which case they could in the meantime take the opportunity to reduce their labour costs. Before the 1760s, local labour shortages in mining could usually be overcome by agreements between colliery owners to lend and borrow those workers who were temporarily unwanted.

Other means of acquiring mineworkers included the poaching of bond labour, contrary to statute law. Widespread illegal recruitment and retention of labour gave rise to many costly disputes and court battles between masters, as did the pursuit and reclamation of runaway colliers. A good example of the fierce and violent nature of such battles over ownership of colliers is indicated in this letter from the Carron Iron Company to their advocate in 1770:

> We lately applied to the Lord Justice Clerk and obtained from him a Justiciary warrant for apprehending upwards of twenty four Colliers who have deserted our works within these three months past. We sent Mr Fish, Overseer of our Collieries with this warrant in quest of these Colliers and he found one of them at Mr Hope's Coalworks at Cowpitts near Inveresk. Mr Fish took with him only one of the Town's Officers. When they were bringing them away the other Colliers belonging to the work set upon them, beat and bruised Mr Fish most unmercifully, deforced the Constable and rescued the Prisoner who made his escape. We do not think we ought to let such an Atrocious Insult pass unpunished and are resolved to prosecute the perpetrators of it ... and apply to the Court of Justiciary for a warrant to apprehend and commit the guilty persons to stand trial. There is one David Gardiner Overseer of the Coal works of Bonharr belonging to Captain Robertson of Earnock who has harboured six of our colliers for some months and not only refuses to deliver them up but threatens vengeance against any persons coming to take them away. They were required from him under form of Instrument on 6th August last. By the Act of Parliament we are informed he is liable in a penalty of £100 Scots for each collier he detains and we intend to pursue him for it. Would you advise us to bring the process before the Sheriff of the County or before the Court of Session?[10]

It is not known to what extent non-servile labour ever existed among the native-born underground workforce in coal mining in Scotland before the abolition of the serfdom laws, but it would appear that the number of such free workers must have been very small before the middle of the eighteenth century. Those colliers who succeeded in escaping from their masters for a year and a day were legally free, but there is no evidence to suggest or to prove that many colliers did manage to abscond and avoid re-bonding. Later, as masters grew more desperate for additional colliers, especially from the 1760s onwards, some offered to engage colliers without binding them. In the 1760s and 1770s, anxious to augment his workforce, Sir James Clerk at Loanhead offered bounties and higher wages to free colliers, and was also accused of poaching free miners who worked at neighbouring Hawthornden. However, throughout the eighteenth century, free labour was attracted and recruited from England, not least in the form of experienced mining engineers and managers, but also skilled oversmen and colliers who were on the yearly bond system which was the norm in England.

Another example from Carron in the 1760s illustrates the remarkable situation where their collier workforce consisted of both bound Scots and newly recruited English free labour. The Company already had leased two local pits, which they worked with native colliers. From the start, however, they were determined to resolve the shortage of experienced and reliable native colliers, and exerted their preference for English colliers who were brought in from Shropshire to work the newly-sunk pits on the longwall system. They were segregated from the Scots colliers, who understandably resented the status and higher wages accorded to the incomers. A set of aggrieved colliers was no recipe for good industrial relations, but the predicament facing the Company in its initial search for reliable skilled colliers also drew attention to the insistent problem of labour shortage.

This bottleneck in the supply of colliers was increasingly becoming a common problem for new merchant capitalists who were seeking to invest in coal mining, but were deterred from going ahead by the scarcity of workers. This problem was particularly acute for prospective business entrants to mining, eager to start working coal on new sites, but confronted with huge problems of recruiting a labour force for the first time.

It can therefore be seen how, in conditions of a rapidly expanding demand for coal in the last decades of the eighteenth century, established and prospective coalmasters alike were facing several accumulating pressures arising from problems related to availability and price of colliery labour. Securing an adequate and suitable supply of workers; having to combat payment of higher wages to satisfy the claims of workers who had learned how to organise to enhance their conditions; and the disruptive and often costly effects of being embroiled in disputes over poaching and desertions, were obvious signs that the existing system of law and practice for recruiting and disciplining collier labour was breaking down and in urgent need of change. The multiple problems facing the coalmasters provided the setting for the ensuing campaigns and debates which were eventually to lead to abolition of the life bonds of serfdom as an intended route out of their difficulties.

The Ending of Serfdom

In considering the motives of the coalmasters and the propertied classes on the issue of ending collier serfdom, it was once thought that humanitarian concern for the plight of this unfortunate class of workers was somehow the driving force behind the Acts of 1775 and 1799 which ended the legal basis of serfdom in Scotland. The campaigns of the 1770s for emancipation of the colliers from serfdom may have been helped by support at that time, within the British Parliament, for abolition of the trade in black slaves. In Scotland, public excitement over the long case of Joseph Knight, a black Jamaican slave whose campaign for freedom was brought against a Fife-based Scottish master before the Court of Session in Edinburgh may also have raised

Plate 18.
Old stoop and room workings.
(*Collection of the Scottish Mining Museum Trust*)

comparisons with collier servitude at home. It is interesting to find that local colliers were among those who raised funds to fight his court case, which was finally settled in his favour in 1778. However, apart from this episode, there is no evidence of any close connection between the debate about the slave trade and the emancipation of Scottish miners. In 1762, emerging from court cases involving disputes over bonding, petitions were issued in the name of aggrieved colliers. They claimed that freedom from the bond would not only secure their own personal liberty, but also ensure that they and other entrants to mining would thereby become willing workers, instead of being reluctant and demoralised ones. Apart from this moderate statement drafted by a friendly lawyer in conjunction with a few informed colliers, between the 1770s and the 1790s there were no reports of popular movement of protest or outcry against collier serfdom on the grounds that it was a deplorable, repugnant and indefensible system in an otherwise enlightened, civilised society. Instead, from the 1760s onwards a few individual coalmasters and informed commentators with enlightened views voiced concern about the immoral and outdated imposition of life-bond serfdom, although their views were mainly overshadowed by more hard-nosed arguments about the economic realities of retaining or abolishing collier serfdom.

Adam Smith, the great economist who came from the mining community of Kirkcaldy, had already explained in 1763 that it was more expensive to keep bound labour than free labour, and argued that, by freeing the colliers and increasing their number, the supposed laws of supply and demand of labour would operate, and masters would be able to cut wages:

> It would certainly be an advantage to the master that they were free. The common wages of a day labourer is between six [2.5p] and eight [3.5p] pence, that of a collier is half a crown [12.5p] If they were free, their prices would fall.[11]

In much the same vein of argument as that voiced on behalf of colliers in 1762, Sir James Clerk, owner of Loanhead colliery, addressing his fellow coalmasters in 1772, stated that serfdom had outlived its usefulness, as it had created a stigma which deterred workers from becoming colliers, and was responsible for their continuing scarcity:

> I have always considered the present state of the laws regarding Coalieries in Scotland as highly subversive of the general interest of the Coall trade and particularly so to the Interest of the Coall maisters.

The servitude the Coaliers have now for more than a century been subjected to has not only been the real cause of the present great scarcity of hands we all justly complain of, but which is infinitely of worse consequence, has raised such a spirit of national prejudice and total aversion among the Inhabitants of this country to that particular business that I am much afraid that even the best regulated schemes we can possibly devise, will prove inefectual, at least for a great number of years, totally to root out these national prejudices … it therefore becomes our duty as well as our interest to promote such schemes as may gradually advance the liberty of the Coalier, and likeways gradually recover the inhabitants of this Country for the prejudices they now ly under, which in the end may procure as many hands to be employed in this valuable trade as in any other.[12]

For Adam Smith and those coalmasters who now seriously questioned the economics of serfdom, it was hoped that an influx of free labour into the pits would also have the ultimate effect of lowering wages. Those who supported the Clerk line of argument recognised the crude reality of removing the stigma of serfdom in order to make the tasks of coal mining sufficiently respectable to attract entrants in the first place.

However, the measures and implementation of the Act of 1775 fulfilled neither of those desires in the immediate or short term. A phased, gradual approach was adopted towards emancipating the colliers, as those coalmasters in favour of change had to placate opposition from coalmasters, principally in the east, who feared that outright, immediate emancipation would result in mass desertions from the pits.

The clauses of the Act are a messy and wordy affair, and can best be summarised to grasp their meaning. The 1775 Act did not immediately free any serving colliers and failed to tackle the continuing bottleneck in the supply of colliery labour. Only new hands entering coal works were to be free and existing mineworkers were to receive their freedom only after a stipulated number of years according to their age. Men in the most productive and main age group between 21 and 35 were compelled to wait ten years before they could claim their freedom Those in the older, 35 to 45 age group could claim their freedom after seven years, providing they trained up at least one young collier apprentice. Women workers and their children were granted freedom with the men. Yet, even after these extended periods of service, freedom was still not automatic, and depended on lodging a formal proof of record of employment service before the sheriff.

It is no wonder that colliers were suspicious of the terms and intentions

behind this legislation, which appeared unsympathetic, obstructive and hostile to the claims of labour. They pondered the prospect of being granted freedom, with some rightly fearing the consequences of being cast out into the open labour market without the social protection they had known while bound to a strict but otherwise not uncaring master. Here, we must remember that life-binding involved mutual obligations under law and custom, however unequal the relationship of master and servant, and the frequency with which many masters evaded such responsibilities. And while no case can be made for the existence of widespread generosity among the class of landed coal owners and coalmasters in late eighteenth-century Scotland, the contract of long service required the master to keep his bondsmen and their families in the basic necessities of life, namely to provide employment and make some provision for them in times of illness, hardship and old age. There was a genuine fear among some of the mining community that the loss of this protective paternalist relationship would result in them being no better than the labouring poor who worked for daily subsistence wages or less, and who had no rights to any other means of assistance for their families than from the parish poor law system, which was seen as a last resort and a shameful fate.

Those colliers who possessed a heightened awareness of such legislation would not have failed to notice the particular, punitive, anti-labour clause inserted into the 1775 Act which added a penalty of two extra years' servitude for participation in a combination against the masters. In other words, freedom from life-binding was conditional upon colliers being able to satisfy local courts that they had not been involved in illegal collective activity. Clearly, against a background of increased labour unrest in the coalfields, with troops being called out against striking colliers at the Carron pits, this was a clause designed to deter and root out militants, and re-impose the authority of employer law.

Some coalmasters made a concerted effort to attract more labour under the new conditions of the 1775 Act. This example from the Carron Iron Company in late 1775, taken from an advert for colliery labour in the Scottish press, indicates its intentions to build a reliable, disciplined and efficient male workforce. Colliers were to be engaged on annual contracts (as in England), or for agreed longer periods; notification given of piece rates and productivity bonuses; and indentured apprentice schemes introduced to train up skilled hewers:

Whereas by the late Act of parliament, no person entering as a collier to any colliery within Scotland can become bound to that colliery,

Carron Company to encourage labouring people to become colliers do hereby give notice that they are ready to enter into agreement with able bodied labouring men and to employ them as colliers for any space that shall be agreed on, not less than one year, and to pay them one shilling per day of wages certain, and whenever they shall have attained such knowledge in their business as that they can work more coal in a day than would cost the Company more than one shilling at the common rate of the field, the Company shall from henceforth pay to such men in place of the certain wages of one shilling per day, the same rates and prices for each ton of coals they shall work that are paid to the other colliers employed in the same kind of work, and a diligent industrious collier can easily earn from eighteen pence to half a crown each day.

The Company also propose to indent young lads from thirteen to eighteen years of age to serve as apprentices to the business of coal hewing, lads of thirteen and fourteen years of age to bind for seven years and to receive ninepence per day for the first three years and one shilling per day for the remainder of their apprenticeship. Lads of fifteen and sixteen years to bind five years and lads of seventeen and eighteen to bind three years, and both receive one shilling per day until any of them can work more coal in a day than to the amount of one shilling. The Company will henceforth pay them by the ton of what they work at the same rates that are paid to the other colliers in the field.[13]

Close scrutiny of the terms of the Act of 1799 which finally ended serfdom and freed all mineworkers reveals clauses which were even more explicit than the 1775 Act in targeting combinations of colliers which, since the early 1790s, had appeared more powerful and widespread, particularly in the west of Scotland. That the principal thrust of the moves leading up to the final framing of the 1799 Act was directed against the power of the colliers' brotherhoods and incipient trade union activity is contained in the very wording of this piece of intended legislation. It was appropriately entitled 'A Bill to prevent combinations among colliers and other persons employed in the collieries, and to regulate their wages and services … in that part of Britain called Scotland'. The wording of this bill went on to condemn the high wages paid to, and extorted by colliers, and their recent habit of working only three days a week. Finally, it called for Justices of the Peace to fix collier wages for a year, and to insist on discharge notes from previous employers when seeking to move to another colliery.

Colliers had not made any evident protest at the passage of the 1775 Act, but their reception to the terms of the bill of 1799 was very different. For the first time, organised collier protest succeeded in mobilising support beyond local workplace and neighbourhood levels, and roused antagonism across the coalfields. Lanarkshire colliers took the lead, around 600 of them raising funds of two shillings each to mount a petition campaign, to employ sympathetic lawyers, and send deputations to other mining districts. Several area petitions were lodged against the terms of the bill, and even coalmasters opposed some of those terms which they feared would inflame further discontent among the collier community. The clauses dealing with discharge notes and official long-term wage fixing were dropped from the wording of the 1799 Act which otherwise contained its extraordinary mixture of abolition of serfdom clauses and disciplinary measures against organised labour.

In an important sense, the 1799 Act can be seen as a coalmasters' charter, instead of a significant turning point in the forward march of labour. The Acts of 1775 and 1799 were placed on the statute book for the benefit of the coalmasters. There is absolutely no way that the abolition of serfdom and the granting of personal freedom to the collier community can be seen as springing from a goodness of heart on the part of the coalmasters and the political elite at Westminster. The final emancipation of the mineworkers was dictated by economic necessity, setting up a free market in the employment of colliery labour so desired by the new breed of aggressive, competitive, coal and ironmasters of west and central Scotland. There is no doubt that the traditional paternalistic estate coalmasters thoroughly exploited their workers and continued to do so into the nineteenth century, especially those who employed large numbers of female pit labour in east and central Scotland. However, as landowners with responsibilities at estate and parish level, there are signs that at least a few of them discharged some obligations towards the basic wellbeing of their colliers and families. In contrast, the more ruthless, single-minded, more capitalist-oriented coal and ironmasters were motivated primarily by business profit, and sought to engage labour on wages contracts without any further obligations towards them. Free labour now meant the freedom for this brand of coalmaster entrepreneur to acquire a labour force in the open market at the lowest possible price without having to contend with costly court disputes and risk having to compensate those owners whose workers they had illegally enticed away. It is interesting to find that the Carron Company, a typical example of this brand of industrial capitalist, did not approve and sponsor sick and benefits clubs amongst their colliers. Instead they regarded such bodies as an encouragement to idleness and absenteeism rather than as self-help friendly societies which provided their workers with a basic safety net against the ravages of industrial accidents, illness and the infirmity of old age.

Beyond Serfdom:
Standards and Experiences into the New Century

The final emancipation of the collier community from serfdom was scarcely more effective than the 1775 Act in encouraging recruitment to the industry. Unskilled workers, including men and families who had been cleared from the Highlands and from Lowland estates, still remained averse to joining an occupational sector which was notoriously dangerous, exhausting and punishing, despite the undoubted opportunities for improved earnings. An unknown number of pit workers used their newly found freedom to leave the industry altogether, and some pit men opted to remain as surface workers, albeit on less wages. Others took employment on enhanced terms of engagement in the rapidly developing mining areas of Ayrshire and Lanarkshire, particularly where the iron industry was creating fresh demand for pit labour.

As piece-rate workers, boosted by their scarcity value and their own collective assertions in bargaining with employers, able hewers could command rising real earnings during the mainly boom years from the 1790s until the end of the Napoleonic Wars in 1815. Then, the wages of an average hewer in the west rose to a peak of around 23/- for a five-day week. Compared to other forms of manual and some skilled labour, the relatively well-paid collier appeared to be better off, while frequently managing a three- and four-day working week. For those colliers who strove to lead respectable lives, and could resist the pressures of excessive drinking and indebtedness, the rising curve of real wages for many colliers had a marked effect on their standard of living and on that of their families and communities. Here, we must take into account the considerable number of wives and daughters who formerly had been bound with their menfolk as underground bearers. Freed from such labour from the late 1770s onwards, some had also exercised their choice to leave pit work altogether and had taken up or resumed roles as homemakers, devoting themselves to family duties and the possibility of creating a more comfortable home environment. Housing accommodation and domestic sanitary standards in mining communities were usually squalid and miserable. Yet, the prospect of more money coming into some households with several male pit workers contributing to the family income, aided by the presence of more women being released from pit work, resulted in a visible increase in dignity, domestic comfort and material possessions.

Those collier families who found themselves in such relatively fortunate material circumstances would probably manage to sustain this improved standard of living for several years, until enforcement of wage cuts took their toll on the workforce from 1816 onwards.

Into the early nineteenth century, coalmasters sought to regain the initiative in contracting and managing their colliery labour. Although the annual binding of free labour was the new norm, some masters tried to persuade men to sign on for longer engagements. This was achieved in most cases by offering the incentive of big signing-on fees, or enhanced bounties. In similar fashion, William Dixon, manager of the large Govan Collieries, introduced a productivity agreement as part of the contract. In 1813, he was requiring colliers on the longwall system of coal extraction to contract to a specific amount of coal for that year (720 carts by agreed weight), for a stipulated rate of pay. This type of minimum agreement evidently suited both employer and hewers. As Dixon commented, the colliers 'know that they can earn good wages, and in a reasonable time bring their engagement to a close by working steadily'.[14]

In the Midlothian coalfield, where the feudal authority of the large landowning coalmasters still predominated, the oppressive system of long bonds over entire families was retained until the 1820s. This meant that the women and children continued to be bound to pit work along with the male hewer. In this respect, it was as if the emancipation legislation of 1799 had never happened. Here, the length of bonds ranged from six months to five years. And although Midlothian had a reputation for the highest rates of collier payment in Scotland, it has to be borne in mind that this was family earnings, including payment to the hewer for the women and child team of underground bearers and haulage workers.

A particularly crude means of enforcing work discipline and inhibiting the mobility of the colliers was to trap them into debt-slavery. Already, before the end of the eighteenth century, some coalowners, coalmasters and their managers were heavily implicated in operating sutleries, where workers were expected to buy goods on credit against future payment of wages. Sutleries were company stores, sometimes also called tommy-shops or truck shops, where the transaction of goods and money gave rise to gross abuse and exploitation. In 1793, Lord Dundonald , coalmaster at Culross, accused fellow coalowners in Scotland,

> or those who pay him an annual bonus for leave to do so ... of the
> most mean practice at many Scots Collieries, of keeping a sutlery, and
> paying the colliers to account of their wages in oatmeal, salt beef, salt

herrings, butter, cheese, soap, candles etc. and together with two
articles although last not least, small Scots twopenny ale and whisky.
In such trash a Scots Collier's wages are frequently spent before the
pay day. The more articles he takes up, the more the sutlery gains,
and which militates against that system of economy which produced
such an alteration on the Colliers at Culross.[15]

In the final words of his condemnation of the truck system, Dundonald is
referring to his own paternalist and well-intentioned, though questionable,
practice of paying his colliers a sufficient proportion of their wages fortnightly
to meet ordinary needs. He was forcing his colliers to adopt a savings
mentality by paying them the balance of their due wages in three or four
instalments within a year. Thus, at Culross, in place of a truck store equipped
with alcohol and shoddy, overpriced articles, his collier families were free to
use other shops as they pleased and were encouraged to be self-respecting
managers of their earnings. This method of paying them, he asserted,

> made them more economical, and the arrears enabled them to
> provide themselves with good clothes, household furniture etc. and to
> lay in a supply of beef for their families in November. Indeed, they
> carried the taste for elegantiorum of life further than may be thought
> necessary; most of them had silver watches, clocks in all their houses;
> several of them on a Sunday wore silk stockings, tambour
> embroidered silk vests, with their hair well dressed and powdered.[16]

In isolated mining communities, with no shops, there is every reason to
believe that the workers initially welcomed employer- or company-led
initiatives to set up a store with a range of basic provisions. However, here
and at company stores in or near coal towns, this essential service too often
gave rise to unscrupulous practices by the employers or their servants. These
exploitative practices included charging excessive prices, selling goods of an
inferior quality, and giving out goods on credit with interest, all liable to be
deducted from the next payout. In the late eighteenth- and early nineteenth-
century context, hard cash in small change was often in short supply in such
locations, and rather than transact shop business in coins, paper slips and
entries in the store-book were used instead to record sales and credit with
each customer. Payment of wages was sometimes handled in the same
fashion, where the pay office and the store were in the same building or next
door; and part of the wages was given to the worker in hard cash, the rest
being entered as credit. It was easy to convert such practices into a system of

abuse, in which colliers and their wives were caught in a vicious cycle of living on tick, their wages further plundered by the extraction of interest. They found it impossible to save, ending up in debt to the store and the master.

By such means, collier indebtedness was deliberately encouraged, and the ready availability of alcohol at the company store proved too strong a temptation for many workers who subsequently found their bargaining position weakened when contract renewal time came round again. Respectable, responsible and sober colliers would do all they could to avoid falling into the clutches of this debt slavery. The truck system was in widespread use across the Scottish coalfields by the early decades of the nineteenth century, notably in the landed estate collieries in Midlothian. It was also prevalent in lead mining communities, where payment of full earnings was delayed for up to two years until the value of market sales of ore had become known. Instead, miners were paid in instalments of cash payments once or twice a year, in such a way as to keep them in debt and remove the option of leaving employment too early for any reason. The company store added to the weight of debt, where credit was obtained for food and drink, tools, candles and gunpowder. At Leadhills and Wanlockhead, the landowners and mining companies established paternalist regimes, mixing strict codes of behaviour with promotion of basic welfare concerns. They encouraged workers to build their own houses, take in waste ground for small-holdings, keep a cow and hold the houses and land rent-

Plate 19.
Leadhills Miners' Library, the oldest subscription library in Britain.

free, with liberty to sell or bequeath to family. They both contributed to a
charity and pension fund for sickness and old age, provided a doctor, a
school and a minister, and in both communities a well-stocked and wide-
ranging circulating Miners' Library, supervised by committees of workers.
According to commentators at the end of the eighteenth century, an
atmosphere of intelligence and sobriety generally prevailed in the two
villages. It seems that the objectives of the masters may have been achieved in
establishing work discipline and moral behaviour. However, for the workers,
any supposed welfare benefits were poor compensation for a low-wage
economy, the company store system and the hardships of work in lead mines.

Among paternalistic coalmasters and new mining companies alike, there
were exceptions, like Dundonald, who did not countenance truck. However,
as will be seen later, systematic operation of truck was to become even more
intensive into the mid nineteenth century, as coal and iron masters, particularly in Lanarkshire, increasingly used it as a weapon to intimidate and
control their mineworkers.

Industrial Conflict: Trials of Strength

In the struggle for control over prices, wages and conditions, apart from
everyday workplace tensions and disputes, masters and workers were
sometimes engaged in fierce confrontations on a new scale up to and beyond
1830. While combining against employers to influence wages and conditions
was an illegal activity, the law did not apply to the masters themselves, who
were free to associate and organise in their own trade interests. Formally
constituted associations of coalowners and coalmasters had not yet emerged
although, in the east of Scotland, groups of masters had met regularly since
the eighteenth century to manipulate and fix the market price of coal. The
coalmasters who supplied the Glasgow trade since the 1790s were also well
organised to protect their interests in the same fashion.

Among the colliers, there was no regular or continuous trade union
organisation at local or regional level until 1816, when a Glasgow and
Clydesdale Association of Operative Colliers was formed. After a successful
three-week strike against wage reductions at collieries around Glasgow, this
union extended into Ayrshire in 1817, covering the coastal collieries and as
far inland as Kilmarnock. Strike combinations were also active in Midlothian
and East Lothian, centred around Musselburgh and Tranent, although they
appear to have been organised separately from those in the west. After bitter
strikes and reprisals, including prosecutions for illegal combination and oath-
taking, all organised union activity was broken up by 1818. Formal organi-

sation and intensive union activity were revived during 1824–5, when an embryonic national network of collier combinations across the Scottish coalfields confronted the coalmasters and the authorities. Highlights of the main issues and episodes in these conflicts are related below.

Combinations and union activity were concentrated on fighting wage reductions, limiting the darg to maintain the value of negotiated rates, and where possible, on operating a closed shop to restrict entry of unskilled outsiders and blackleg labour into the pits. From the records which were confiscated by the authorities, we know that the collier unions in Glasgow, Lanarkshire and Ayrshire in 1816–17 had over two thousand subscribing members, and a considerable fund for organising committee and delegate meetings, and paying out strike money. This commitment and funding enabled the union to sustain striking members for long periods and to win some successes in preventing wage cuts and enforcing closed shops at individual collieries. In November 1817, the Glasgow committee supported the longest spell of industrial action known until then among colliers – a ten-week strike at William Dixon's pits at Greenend and Calder, near Coatbridge, before it was defeated by the arrest of leading members. At Ayr colliery, also in 1817, the strike was defeated by an intake of Irish labour, perhaps the first of many occasions in which this source of immigrant labour was deployed to break collier unity and inflict lower standards of pay and conditions.

Midlothian colliers had revived union action by 1823, funds being paid out to support striking miners for several weeks. Pressure to drift back to work was met by desperate violence against blacklegs, including the first known cases of ear-cropping as punishment for breaking agreed restrictive practices and the oath of collective solidarity.

Following repeal of the Combination Laws in 1824, conceding a legal status to trade unions, colliers were among craft groups who took the opportunity to renew the building of formal organisation. Articles of the Ayrshire Colliers' Association declared their intent, concentrating on wages and on control of entry into the occupation of collier. The wording of the article reproduced below on apprenticeship regulations and conditions of entry for outsiders was similar to that endorsed by the other collier associations in the west and by the federation of Associated Colliers of Scotland in 1825. By setting precise age limits, the output of young workers was to be restricted, and full hewers were prevented from increasing their allocated darg by taking on boys to assist them. Preference is given to the sons of colliers, accompanied by high fees and the imposition of more stringent conditions on outside entrants. Significantly, there is no place for underground female workers in this proposed scheme of workplace control, as the unions of independent colliers had taken a stand against the employment of female labour, primarily on the grounds that this practice tended to encourage and enable masters to lower the overall earnings from pit work.

> Article VI. Any operative collier having a son wishing to go to the trade, at the age of ten years, will be entitled to one fourth of a man's work; at thirteen years of age to one half; at fifteen to three fourths; at seventeen to full work ... Any person coming into the trade who is not a collier's son must serve a regular term of three years, and pay five pounds sterling entry money; one pound to be paid each year, the whole of the monies into the Association, to find security for the payment ... He may be entitled to half a man's work at sixteen years of age; and to three quarters at seventeen; and being clear, and all his money paid up, may have full liberty at nineteen years of age. No neutral man can engage to become a collier without paying seven pounds, and serving three years ...
>
> Article XI. Is it not evident that there are masters in the coal trade who are constantly running a race in the deduction of wages, and are never satisfied unless they are paying below their neighbours, and by forcing the measure far above the common standard to find a sale and outsell their neighbour colliers?. This is a case which requires immediate attention, and it becomes the duty of this association to point out such masters, and after being duly warned, if they still continue in such a career, then it will be our duty to put them out of the trade.[17]

These strong words were put to the test in 1825. George Taylor, owner of Ayr colliery, who had broken the 1817 strike with blackleg labour and continued to take a firm anti-union stand, was one of the chosen union targets. Buoyed up by successful limitation of the darg at other collieries, union men attempted to appoint themselves as checkweighmen at the mouth of the winding shaft, to keep account of the hutches and loads as they came up and to estimate the darg done by each hewer. This interference with managerial authority was resisted and a strike ensued, whereupon Taylor again managed to recruit a squad of strike-breakers, principally Irishmen, who were given protection, quickly installed and inducted by a few loyal skilled colliers.

A series of protracted confrontations in other parts of the Scottish coalfield during 1825 and 1826 proved major tests for local and wider trade union solidarity and for counter-offensive action by employers and authorities. The first and most prolonged contest was at the Redding and Brighton collieries, along the new Union Canal near Falkirk, where the owner, the Duke of Hamilton, took a determined stand against wage increases and industrial action. Two hundred colliers stayed on strike for nearly five months from December 1824 until April 1825, sustained by funds poured in from the Associated Colliers of Scotland. Strike levies were raised from all over the coalfields, from Ayrshire to Fife.

Intent on breaking the strike and re-starting production, the Duke ordered up a large squad of miners and labourers from his estates in Hamilton. They were ambushed close on arrival and, faced later on Redding Moor by an intimidating demonstration of nearly 700 colliers armed with cudgels, they wilted and were prepared to start work only under armed guard. Meanwhile, leading colliers were pursued and arrested by sheriff and posse, sent for trial and subsequently imprisoned for up to a year on charges of assault and intimidation. Eventually, the strike crumbled as the Hamilton squad continued to produce coal and a gradual drift back to work began among the established workforce.

While the failure of the long strike action at Redding was a setback for the colliers' union, fellow members maintained their offensive in Lanarkshire, Renfrew and Glasgow. In June 1825, concerted rolling strike action brought out all the men at the collieries of Colin Dunlop, owner of the Clyde Iron Works, near Cambuslang, Although Dunlop evicted strikers from the company-owned houses, the colliers stayed firm, all 300 maintained with ten shillings a week in strike funds. His ironworks at a stop, Dunlop was forced to accede to the colliers' demands.

This success was followed by widespread enforcement of output restriction at many collieries as far east as Fife, and by autumn 1825 the leading coalmasters were united in their determination to stage a counter-offensive and destroy the power of the colliers' unions. Where Dunlop failed meantime to contain the colliers, the more formidable coal and iron master William Dixon refused wage demands at his Faskine pits near Airdrie, evicted all the strikers and took on new workers under protection of armed guards. When union delegates attempted to bring out the men from Dixon's principal Govan collieries in Glasgow, he alerted the authorities to arrest them, resulting in two-month jail sentences for restraint of trade, i.e., interfering with the employer's right to run his own business.

Dixon's hard-line resistance proved a turning point and an example to other masters. In response to worker demands, Lord Belhaven, in the Wishaw area, the owners of Muirkirk Iron Works and the large Hurlet Coal and Lime Works in Renfrewshire, all dismissed their mineworkers and recruited a fresh intake on the masters' conditions. The final showdown was fought out, again, at Colin Dunlop's Clyde Iron Works in April 1826, by which time the trade boom had ended, prompting further attempts at wage reduction. Here, industrial warfare followed the usual pattern, with strike action, eviction from company houses, replacement of labour under guard and guerrilla action over several weeks between pickets and colliers wanting to return to work. Defeat and disintegration followed for the Lanarkshire colliers' union, as masters dug in elsewhere, replaced their workforce, set new rates and re-established their authority.

Vital planks of the colliers' strategy – to limit the darg and to control entry to the pits – had been remarkably successful during the previous trade boom, but were seen to have failed their members in times of depressed trade. However, other hard lessons were in store, as influential ironmasters had been forced to conduct the struggle to the finish against the colliers' offensive. As consumers of their own coal, they had to absorb increased labour costs in order to maintain a competitive price of pig iron in the open market and consequently avoid significant reductions in their profit margins. A sharp decline in the price of iron in 1826, as on many future occasions, meant almost inevitable conflict with their workers over wages and conditions. For those ironmasters whose collieries supplied coal for their furnaces, the strategy from now on was to defeat collier pretensions to assert controls in the workplace. To do this, they had begun to identify the need to build up their own nursery of colliers and oncost workers from a core of reliable experienced men who could train up new recruits, including unskilled outsiders, for their respective work tasks. This would take time and patience, and not always yield the best results in terms of productivity in the short term. However, persistent employers found that strike-breakers and fresh labour could be employed successfully in those pits which had thick seams, where hewers and teams could extract coal which was more accessible than normal conditions permitted. This included pits in which the longwall method of coal-getting was practised, as at Dixon's Govan and Calder collieries, and where the job of coal winning could be learned more quickly under closer supervision than was ever possible in those pits where the colliers worked alone or in pairs. Moreover, new entrants were also prepared to accept lower wages than the independent collier, especially while they were being trained up to their responsibilities of hewing and hauling coal.

A reserve pool of unemployed and displaced workers, such as handloom weavers, and an influx of Irish labouring poor, all desperate for regular paid work, had become available after 1815. Indeed, by 1826, when Irishmen

were used to break the strike at Clyde Iron Works, coalmasters were able to overcome the longstanding labour shortage, while re-asserting the upper hand in the workplace. This successful challenge to the status of the independent collier in the trial of strength between masters and men between 1816 and 1826 resulted in the defeat and collapse of this pioneering episode of trade union organisation across the Scottish coalfield. It also hinted that, into the indefinite future, collier resistance to workplace grievances and to the authoritarian stance of the masters would be localised, sporadic and possibly violent.

Profiles of Working Lives in Mining, c. 1700–1840

Work and Working Conditions

In surveying the workings of an extensive colliery below ground, a married woman came forward, groaning under an excessive weight of coals, trembling in every nerve, and almost unable to keep her knees from sinking under her. On coming up, she said in a most plaintive and melancholy voice: 'O sir, this is sore, sore work. I wish to God that the first woman who tried to bear coals had broke her back, and none would have tried it again.'[1]

No.1 Janet Cumming, 11 years old, bears coals:

Works with father, has done so for two years. Father gangs at two in the morning; I gang with the women at five, and come up at five at night; work all day on Fridays, and come away at twelve in the day.

I carry the large bits of coal from the wall-face to the pit bottom, and the small pieces called chows, in a creel; the weight is usually a hundredweight; does not know how many pounds there are in a hundredweight, but it is some work to carry; it takes three journeys to fill a tub of 4 cwt. The distance varies, as the work is not always on the same wall; sometimes 150 fathom, whiles 250. The roof is very low; I have to bend my backs and legs, and the water comes frequently up to the calves of my legs; has no likening for the work; father makes me like it; mother did carry coal, she is not needed now, as sisters and brothers work on father's and uncle's account. Never got hurt, but often obliged to scramble out when bad air was in the pit.[2]

When the nature of this horrible labour is taken into consideration, its extreme severity, its regular duration of from 12 to 14 hours daily, which, once a week at least, as in the instance of Janet Cumming, is extended through the whole of the night; the damp, heated and unwholesome atmosphere in which the work is carried on; the tender

Plate 20.
Janet Cumming.

age and sex of the workers; when it is considered that such labour is performed not in isolated instances selected to excite compassion, but that it may be truly regarded as the type of the everyday existence of hundreds of our fellow creatures – a picture is presented of deadly physical oppression and systematic slavery, of which I conscientiously believe no-one unacquainted with such facts would credit the existence in the British dominions.[3]

This chapter is intended as a companion piece to the previous sections of the book. In the opening chapters, the work process and working conditions of colliers and of other members of the labour force in mining have been covered briefly, while attention has been concentrated on other essential themes. So far, the foundations, uneven growth and development of mining, the status and experience of the mining community in the years of serfdom, and the early struggles of the colliers against their landowner and company masters on wages-and-conditions issues have occupied the foreground of our narrative. Now, attention is focused on describing and analysing in much fuller detail the nature of the tasks done by the various sections of the workforce in time and place, including the labour of children, women and adult males, both above and below ground.

The time-span of this focus on the work process and the composition of the workforce is the long formative period from the eighteenth century to the 1840s, before the decisive transition to the development of deep mining, modern technology and the employment of a male-only labour force underground. This part of the study of working lives also allows us to encompass the startling revelations of the famous Children's Employment Commission enquiry and report of 1841–2, which includes abundant and sensational graphic evidence on contemporary and earlier working conditions in Scottish mining. It is an invaluable primary source, especially for its detailed coverage of the labour of women and children in the coal mines of the East of Scotland, where this practice was most prevalent. Sub-Commissioner Franks, who conducted the investigation in the East of Scotland, covered over 100 collieries in the Lothians, Fife, Stirlingshire and Clackmannan. In the process, this thorough, high-calibre social investigator conducted over 400 interviews, the majority with child workers, which was his principal remit. This far-reaching parliamentary enquiry into working conditions in mining was a vital prelude to the historic legislation of 1842, which prohibited the employment underground of women and girls, and of boys below the age of ten. For these reasons, it is appropriate that the richly documented evidence and testimony provided by the two officially appointed sub-commissioners who investigated the Scottish scene and the many workers who they interviewed is fully represented in this overall profile of mine work.

The selection of extracts from the 1842 Report, from other contemporary sources, and the accompanying discussion, are intended to reflect and highlight working life along the following lines: the different composition of labour across the geographical extent of the Scottish coalfield, the age and gender of the workforce, the division of labour above and below ground and any other features, not previously covered, that may be regarded as being distinctive or peculiar to the work environment of mining in eighteenth- and early nineteenth-century Scotland.

Male Colliers

The first case study is the hewer, as the principal production worker among the male colliers below ground. First-hand accounts of the work of the hewer for this period are rare from Scotland, although knowledgeable observers provide a useful impression:

> The art of the collier is to hew down immense blocks of coal, in size from one to four cubic yards at a time. The next step is to divide this block into pieces. … In working the Scotch coal, which is very strong in the wall, it requires such constant exertion and twisting of the body that, unless a person have been habituated to it from his earliest years, he cannot submit to the operation. For instance, it is a common practice for a collier, when making a horizontal cut in that part of the coal which is upon a level with his feet, to sit down and place his right shoulder upon the inside of his right knee; in this posture he will work long, and with good effect. At other times, he sits with his body half inclined to the one side, or stretched out at his whole length, in seams of coal not thirty inches thick.[4]

A clear account of the main task of the hewer is provided by Thomas Tancred, the sub-commissioner for the West of Scotland, as part of his report for the Children's Employment Commission. He reaffirms Bald's earlier description of the physical stance and rhythm of the work process as adopted by the hewer:

> In whatever way the coal is worked, the labour of the collier is one of the hardest of which I am acquainted. The thickness of the seam sometimes affords him more space to work in than is the case in the

generality of pits, but yet he seldom stands to his work. The ordinary posture is sitting with one leg doubled beneath him and the other foot resting against the coal. Inclining his body to one side so as to nearly touch the ground with one shoulder, he digs his pick with both hands into the lower part of the coal, or haply into a stratum of fireclay or some softer material beneath the coal. In this way he picks out an excavation often for a considerable distance under the mass of coal, beneath which he half lies to work. When he has, after two or three hours labour, undermined as much as he judges it prudent to attempt, he inserts iron wedges by means of a heavy hammer between the coal and the roof above it by which, and by the weight of the ground above, the mass of the coal is detached and falls. The cramped posture, the closeness of the subterranean atmosphere, loaded with coal dust and the smoke of his lamp, and sometimes with sulphurous exhalations, together with the bodily exertion, cannot fail to be very exhausting.[5]

This description of coal extraction was typical of the stoop and room method and the workplace, whereby each hewer worked in a room or chamber cut into the coal seam. These work-spaces were separated by stoops, or pillars of coal which supported the weight of the roof and were left unworked. The dimensions of the rooms and the stoops varied, with judgement being determined by the thickness of the seam, the firmness of the strata

Plate 21.

Boy hewer with pick and tallow lamp.

immediately above and below it and the depth of the workings. In 1841, Thomas Tancred found rooms typically 12 feet wide and stoops 15 feet square in the West of Scotland mines, to a depth of 200 feet. However, elsewhere in the area, in a minority of deep workings to around 400 feet, huge stoops of up to 40 feet square were being left to support the roof.

Usually, one or two adult hewers would work in a room, assisted by a boy hewer for coal cutting and other tasks such as filling and perhaps drawing hutches. In some situations, adult hewers themselves also filled the hutches and, if the roadways were sufficiently wide and high, did their own haulage to the pit bottom. The age of the boy hewer could range from nine or ten to seventeen, by which time the lad could have served an informal apprenticeship in learning the practical tasks of a collier, and become eligible for a full place, or 'ben', in his own right. Hewers had an agreed notion of the stages in the right of passage to full hewer. This notion has already been noted earlier in connection with the articles of the colliers' trade unions in the 1817–25 period. And although colliers were not always able to exert control over the employment and allocation of tasks to boy assistants, the idea still existed as a powerful aspiration in the minds of the independent collier fraternity. Andrew Miller, a mine manager in the Monklands area, and author of an important pioneering book published in the 1860s on the history of Coatbridge, looked back on this issue for the earlier years of the nineteenth century:

> The old Scottish colliers looked upon their profession as a sort of hereditary right ... and for which they had to undergo a regular apprenticeship. At the age of eight or nine the boy was sent to work in the pit as a trapper, where he opened and shut a door which aided in the ventilation of the mine. In a couple of years he became a putter, and assisted in pushing the loads of coal from the workmen to the pit bottom. In a couple of years more he was termed a 'half man', when he assisted the working colliers, and after another two or three years he was entitled to rank as a 'man' and invested with full powers to wield the pick, wedge and mallet at the coal wa'. Thus, step by step, he was trained to the work.[6]

Alexander McDonald, leader of the Scottish miners, giving evidence to a Parliamentary committee in 1866, also recalled the stages of passage from boy worker to pitman in the early years of the century. At each stage, the youth was allowed to work a proportion of a full man's darg, according to his growing status. Thus boys 'were a quarter of a man from ten to twelve and a half; they were half a man from twelve and a half to fifteen; and at seventeen obtained full man's rights.'[7]

The 1842 Report remarked of 'the early age at which the workers begin to get or hew the coal. It is scarcely to be credited, and the evidence proves, that this labour is performed by male children from nine years and upwards.'[8] John Allen, a 12-year-old, had already worked at the Plean pit, near Stirling, for two years, alongside his father. The boy complained of being constantly tired, and worked in a wet pit, 'which is full of water; the water having risen above the dip'.[9] Robert Bowman, a 10-year-old hewer, 'was learning to hew the coals with father' at Donibristle pit near Aberdour.[10]

Alex Reid, aged twelve, was doubling up as coal hewer, then as filler and haulage drawer at the large Sheriffhall colliery, near Edinburgh, working up to fifteen and sixteen hours a day inclusive in the worst of conditions. Commissioner Franks reported his testimony, as follows:

> I have worked two years at Sheriffhall, and go below at two or three in the morning, and hew til six at night; after that I fill and put the carts on the rails to pit bottom. ... The pit I work in is very wet; we often work in slush over our shoe tops. When first below, I used to fall asleep; am kept awake now. It is most terrible work; I am wrought in a thirty inch seam, and am obliged to twist myself up to work on my side; this is every-day work except Friday, when I go down at twelve at night, and come up at twelve at noon.[11]

The final example in this profile of boy hewers is 16-year-old George Reid, who also worked at Sheriffhall:

> I pick the coal at the wall-face, and seldom do other work; have done so for six years; the seam is twenty-six inches high, and when I pick I am obliged to twist myself up; the men who work in this seam lie on their broadsides.
>
> Father took me down early to prevent me from going o'erwild about the town; it is horrible sore work; none ever come up to meals. Pieces of bread are taken down; boys and girls sometimes drink the water below, when there is no metal in it; men take a bottle of small beer.
>
> I should not care about the work if we had not so much of it; have often been hurt; was off idle a short time ago, the pick having torn my flesh while ascending the shaft.[12]

From this and other accounts, we know that 'born and bred' hewers and colliers who were initiated into the work at an early age and survived the working environment into adult life were most likely to become the best workers. As a mine manager with a special interest in recruiting able colliers, Robert Bald, writing in 1808, was certain 'that the number of colliers cannot be increased at pleasure, as can be done with mechanics and labourers; the latter can begin at any time, when young, and from any class of society; but he who is to become a Scotch collier, must begin his labour as soon as he is able to creep to the coal pit.'[13]

The possession of physical strength, stamina and a supple body were essential characteristics of the able hewer, but the highly skilled craftsman hewer had to exercise other qualities in his everyday work below ground. To protect his own life and that of others around him, he had to apply independent judgement and vigilance to sustain the safest possible immediate working environment. He had to apply a range of learned skills, drawn from practical observation and experience. Such skills included efficient and safe undercutting and holing of the coal seam, sometimes using explosives for shot-firing; a practical knowledge of the coal wall and its mineral properties, so as to distinguish between different types and qualities of coal; recognising rock hardness and faults in the strata; listening to and interpreting the sounds of the coalface and roof; assessing the need for timber propping and carrying out that work; and acting as the fireman, or helping him to detect early warnings of gas and water levels. Beyond those abilities, the best colliers also possessed something akin to an intuitive sixth sense, which arose from long experience of and insight into the many dangers of the pit.

Given all, or even a few, of those necessary qualities of character and skill, we have to agree with a leading historian that it is ludicrous to conceive of earlier situations, during collier serfdom, whereby beggars, vagabonds or common criminals could be taken underground, be forced to work in the dangerous conditions of stoop and room coal cutting, and somehow not be especially prone to make the kind of mistakes that could bring premature death to themselves and the regular workforce. It is impossible to see such men become hewers, in anything approaching the proper sense of that term, but to envisage them instead as pit labourers, which is what most of them were and remained.[14]

The variety of 'graith', or hand tools required by the hewer, is another indication of his range of work skills. A choice of picks was necessary for different tasks of holing and undercutting; a hammer and wedge prised down the undermined coal; a sledge-hammer was used for breaking up very large lumps; drills, chisels and scrapers were required for holing; rods and fuses for blasting; shovels for filling hutches; and perhaps axes and saws for finishing props. This large tool kit was a heavy load, and was usually left at a safe place underground.

Furthermore, the experienced, versatile and accomplished collier was a skilled journeyman and pitman, and an 'aristocrat of labour', although he had not served the formal apprenticeship of a skilled artisan or tradesman. In the many smaller collieries in Scotland during this period, with less than twenty of a workforce, the male colliers were the key production workers, mostly employed in their principal task as hewers. However, due to the limited division of labour available for other tasks that required strength and expertise, the most experienced productive hewers were also frequently called upon to perform those additional responsible and specialist tasks, such as underground road maintenance and the construction of new levels and pit roads. Such hewers were contracted periodically on day- or task-related rates to sink new pits, cut air courses or tunnel new levels, under the supervision of a mine manager or an outside viewer or consultant.

Without the aid of heavy-duty and precision power machinery, sinking and cutting wet levels was perhaps the hardest and most uncomfortable work for the skilled collier. Robert Bald, in his day perhaps the most experienced manager, consultant and authority on practical mining in Scotland had this to say about the nature of the task involved in sinking new pits. It was:

one of the most laborious, wet and dangerous employments that can be imagined, particularly the sinking of engine-pits, where the smoke of the gunpowder prevents the rays of light from penetrating; so that the workmen are enveloped in the most terrible darkness, where the glimmering light of a candle only extends a few inches, and just serves to show how impenetrable the darkness is. While employed at this work, there is nothing to be heard but the clanking and rebounding of machinery, the impetuous rushing of water, and the re-echoing sound of the ponderous hammer, while every other hour they have to lay a train to gunpowder; and quickly springing to the basket, are drawn up the pit by the aid of machinery, with great velocity to escape being blown to pieces; and, it frequently happens that the train takes fire ere they have descended a few fathoms; so that the splinters of stone fly around them in all directions; and the sound of the explosion is so overpowering as to make ears tingle and suspend the sense of hearing for some minutes; yet in the midst of these hairbreadth escapes, they go on with their work in the most persevering style.[15]

Around 1840, Alexander McDonald, while a young but experienced pitman, took on this kind of arduous, better-paid work, to raise extra money to pay his entrance fees to Glasgow University:

> I followed a class of work which we call stonework, that is working all wet hard work, not only working coal or ironstone. I took that class of work because I had an object in view and in that way I could make more money than I could have done in working coal or iron. To give you an instance of the kind of contract I had once, I drove a level of forty fathoms down, going under water. It was wet in the bottom and there was wet pouring down from the roof. This was an ironstone mine and water coming out of the cutters or joints of the roof, and in the morning when I entered the first thing that I did was to roll myself all my height in the water and wet every part of the body for the express purpose that the water falling from the roof should not create the very unpleasant sensation which water does to one who has gone recently from his bed into the mine that is wet.[16]

The arrangement and conditions of the work tasks of the hewer in the emerging longwall system, as practised in some collieries in Scotland in the late eighteenth and early nineteenth century, were different from those of the hewers who worked within the more usual stoop and room system. For example, the independent collier who worked at the coalface in the chamber or room was relatively isolated, unsupervised and had some control over his work effort. In longwall working, hewers were part of a team who worked together in a coordinated fashion to extract agreed quotas of coal under the direct supervision of a subcontractor or an oversman.

At the Carron collieries, the method of longwall extraction had been introduced by miners brought from Shropshire (hence it became known as the 'Shropshire system'). This technique spread to neighbouring West Lothian deep mines, and was being practised in lower Lanarkshire by the 1770s and in the large Govan collieries, Glasgow, before 1800.

Carron had set the pace with new work organisation in the longwall version. Colliers operated in two specialised groups in morning and back shifts. The first shift holed and undermined the lower part of the thick seam with picks, while the back shift attacked the upper seam with crowbars, wedges and hammers, producing various sizes and types of coal. By this procedure, a broad length of a seam was worked, extracting all the available coal. The roof of the cleared area was supported by the accumulating waste of small coal and stone, and with temporary wooden props. Shifting coal waste into the cleared space, and furnishing, erecting and removing props required extra labour from the collier team.

A variant of this technique was used at Govan, as witnessed by Tancred in 1841. On their shift, the hewers advanced along the coal wall, each man working a small section. In the following shift, it was the task of oncost men, 'called rippers, to blow away the roof where the coal has been worked out to about 7 feet high from the floor, and with the stones thus produced, others called the redsmen, build walls 10 feet thick for the support of the roof, leaving 6 feet in width between the walls for road-ways where required. This oncost work is always done during the night.'[17]

Ironstone mining was also carried out by a small team of hewers on a modified longwall basis. As ironstone was frequently found in thin narrow seams, the working space for cutting was restricted in such conditions. Hewing and hauling ironstone were both considered heavier work than the identical tasks in coal mining. Again, according to Tancred:

> The labour of ironstone miners is often worse than that of colliers. I have seen them at work in a space of from 22 inches to two feet high, where even when seated a man could not keep his neck straight, and to get into the place where he was at work was no easy matter to me. The management of his heavy tools in such a confined space must be very fatiguing. Two men take between them 14 yards of the band of stone, and make their own walls of the roof which comes down when the stone is extracted, leaving a road six feet wide to every space of 14 yards.[18]

At the Crofthead ironstone pits, near Whitburn, the contractor, John Reid, in his evidence to the Commission, supplemented this account of the hewers doing their own back filling: 'Men who are wrought upon iron-stone make their own roads, which rarely exceed 54 inches high, and, from the softness of the roof, soon settles down to 40 inches. The whole of the iron-stone is wrought out, and the walls, or support, made up of the waste, on the long-wall system'[19]

Down the Mine: Women and Girls

Precise data is not available for calculating the number, age, gender and occupational categories of mineworkers in Scotland during the eighteenth and early nineteenth centuries. However, despite having to rely on approximate numbers in most cases, the weight of evidence, combined with graphic descriptions of pit work, makes a grim impression. In particular, the employment of women and young children underground, especially in the east of Scotland in the long period before 1842, confirms that harsh picture.

In 1842, the summary findings of the Children's Employment Commission produced some stark statistics about the composition of the labour force in mining within Scotland. They concluded that female employment in mining throughout the British coalfields was found mainly in the east of Scotland. Almost half of the estimated total number of women and girls employed underground at that time were found there, amounting to over 2,300 out of 5,000 in Britain as a whole. Other areas where women and girls were employed underground were West Lancashire, West Yorkshire, and the Monmouth and East Glamorgan parts of the Welsh coalfield. In the large Durham coalfield, female employment in pit work had ceased since the end of the eighteenth century, and in the Cumberland mines had been abandoned some time before 1841.

Moreover, the Commission found that the 'bearing system' – women and girls carrying coal on their backs – was peculiar to the Midlothian and East Lothian coalfields, but was known to be in existence until recently in all other parts of the east of Scotland coalfields from Fife to the eastern borders of Lanarkshire. There are scattered references to the existence of female labour as bearers in stair pits in Ayrshire and in central Lanarkshire during

Plate 22.
Female coal bearer.

the eighteenth century. The evidence is unclear about the extent of the practice in those areas, but numbers are likely to have been small. According to Robert Bald in 1808, coal bearing had been 'abolished in the neighbourhood of Glasgow', and, it may be added, elsewhere in the western coalfield.

On the basis of the figures supplied by the east of Scotland colliery owners in 1841, which were not quite complete but, according to the Commission, 'as accurate as can well be expected', the total number of 2,341 female pit workers divided into 1,189 women and 1,152 girls under eighteen. It is not possible to work out how many of those women and girls were bearers or were otherwise employed underground in the Midlothian and East Lothian pits, but the bulk of the 350 women listed for the Midlothian collieries alone are likely to have been bearers, while the girls under eighteen would have been employed as bearers or as other haulage workers, pushing or drawing filled sledges or trucks.

Females had long been employed as bearers for husbands, fathers and other colliers. This practice can be seen as a custom dating back to the earliest days of mining in the east, when the wife carried out the man's coal as part of her wifely duty. Historically, it was also a survival of the notion that it was an insult to the man's dignity to carry out his own wrought coal. In the improved agricultural areas of the rural lowlands, as in mining, a similar custom existed, whereby the male farm servant (the hind) was hired by the tenant farmer on condition that he could supply additional labour from the female members of his family, or take in a woman or girl (bondager) to carry out the many unskilled jobs throughout the farming year.

At many of the East of Scotland coal mines the bearers outnumbered the men and boys. For instance, at Loanhead, throughout the eighteenth century, the bearers were almost twice as numerous as the hewers; at Pitfirrane Colliery, in Fife, 22 female bearers worked for 16 colliers in 1777; and at the Bo'ness mines in the 1760s there were 102 bearers to 53 colliers. The evidence would seem to suggest that a collier kept two bearers, usually a wife and daughter, and perhaps shared a third with another collier, according to their availability and to cope with any added amounts of coal output.

Before 1842, in the West of Scotland coalfield, very few females were allowed down the pits to work. The obvious exception was in some coal and ironstone pits in the eastern part of Lanarkshire, where small numbers of female workers were still to be found. Here, as will be shown later, girls and women were employed mostly in hauling or pushing trucks filled with ironstone and fireclay.

Analysing the figures from the 1842 Commission Report, there was also a notable age differentiation in the mining workforce between the east and west of Scotland. In the west, around 25 per cent of the young men and boys were under eighteen years of age, and 10 per cent under thirteen. In the east, with a total of nearly 10,000 mineworkers, the picture was very different. It showed a workforce of 40 per cent teenagers of both sexes and almost 25 per cent (2,200 was the estimated figure given by Franks) who were under thirteen years of age. Here, the workforce was predominantly young, whereas in the west the workforce was mainly adult, although boys were also prominent.

Female Coal Bearers

The engagement and working conditions of coal bearers of all ages was a shameful indictment of wilful and needless exploitation, involving and implicating some of the most powerful and wealthy property and mine owners in the land, who were in the best position to end such practices. At another level, evidence points to the attitudes among collier families themselves. Enforced ignorance, the strength of custom, male dominance and the sheer need to make up a sufficient family wage, all helped to perpetuate this degrading, oppressive and abusive system of putting women and girls to work as coal bearers to earn a subsistence living. The respectable, independent colliers would not take their wife or daughters down the pit, and could afford to take this decision, but they were in a minority among colliers at this time.

The chapter opened with extracts illustrating the experience of women and girl bearers who carried creels or boxes of coal on their backs from the coalface to the pit bottom, or even further, up to the surface. These images are now followed and reinforced by additional accounts of the time.

The first account, by Robert Bald, in 1808, and reproduced at some length, is recognised as a classic statement of its kind. He started his enquiry by noting the contemporary methods of hauling coal from the working face to the pit bottom and to the pit-head. The most approved method was in those pits with sufficiently high and wide roads to allow horses to pull loaded hutches to the pit bottom, before the hutches were then hoisted by gin or steam-powered winding gear to the surface. The second method, where the space of pit roads was more confined, involved the pushing or pulling of small wheeled wagons by the collier himself, or by other men, boys or women. The final method was by bearing coal on the backs of women and girls, either to the pit bottom, or up to the top. This, for Bald, was 'the most severe and slavish' mode of carrying coal:

We conceive it proper to bring into view the condition of a class in the community, intimately connected with the coal-trade, who endure a slavery scarcely tolerated in the ages of darkness and of barbarism. The class alluded to is that of the women who carry coals under ground, in Scotland – known by the name of Bearers. Let us now take a view of this system, the severity of the labour, and the consequences.

In those collieries where this mode is in practice, the collier leaves his house for the pit about eleven o'clock at night, (attended by his sons if he has any sufficiently old), when the rest of mankind are retiring to rest. Their first work is to prepare coals, by hewing them down from the wall. In about three hours after, his wife, (attended by her daughters, if she has any sufficiently grown) sets out for the pit, having previously wrapped her infant child in a blanket, and left it to the care of an old woman … The mother having thus disposed of her younger children, descends the pit with her older daughters, when each, having a basket of a suitable form, lays it down, and into it the large coals are rolled; and such is the weight carried, that it frequently takes two men to lift the burden upon their backs: the girls are loaded according to their strength. The mother sets out first, carrying a lighted candle in her teeth; the girls follow, and in this manner they proceed to the pit bottom, and with weary steps and slow, ascend the stairs, halting occasionally to draw breath, till they arrive at the hill or pit top, where the coals are laid down for sale; and in this manner they go for eight or ten hours before resting. It is no uncommon thing to see them, when ascending the pit, weeping most bitterly, from the excessive severity of the labour; but the instant they have laid down their burden on the hill, they resume their cheerfulness, and return down the pit singing.

The execution of work performed by a stout woman in this way is beyond conception. For instance, we have seen a woman take on a load of 170 lbs, travel with this 150 yards up the slope of the coal below ground, ascend a pit by stairs 100 feet, and travel on the hill 20 yards more. … All this she will perform no less than twenty four times as a day's work.

The weight of coals thus brought to the pit top by a woman in a day, amounts to 4080 pounds, or above thirty six hundredweight, and there have been frequent instances of two tons carried.

To many who may read this account of work, the amount of it will not be very obvious, because the depth of the pit and slope of the coal, compared with the same horizontal distance above ground, appear no very great matter.

But in order to make the amount of this work to a standard or scale by which it may be compared, we shall take, for example, a well known and familiar object, the steeple of St Giles, Edinburgh, the height of which is 161 feet from the street to the weathercock.

Now let us suppose that a scale stair were carried up from the base of the steeple to this height, and a platform made there, and that thirty six hundredweight of coals were laid down at the distance of 150 yards from the base – a coal-bearer would make twenty four journeys to this great height, and lay down upon the platform the whole quantity of coals ... In short, the height ascended by them when loaded, is equal to more than four times that of Arthur's Seat above the level of the sea, or to the height of Ben Lawers in Perthshire, the total ascent being 3672 feet. ...

She would perform the same work, five days each week, and that not for a week, a month or a year, but for years together.

From this view of the work performed by bearers in Scotland some faint idea may be formed of the slavery and severity of the toil, particularly when it is considered that they are entered to this work when seven years of age, and frequently continue till they are upwards of fifty, or even sixty years old.[20]

Over thirty years later, in 1841, R. H. Franks was both surprised and horrified to find a girl, not yet aged seven, as a coal bearer. This was at Harlaw Muir and Coaly Burn, small, badly ventilated, country pits in West Linton parish on the fringe of the Midlothian coalfield. Here, the coal-mining lessee was a clergyman, the Reverend J. J. Beresford, who employed fifty mineworkers, among whom were twelve girls well under thirteen years of age. Franks described the six-year-old Margaret Leveston, the youngest child mineworker he interviewed, as 'a most interesting child, and perfectly beautiful' and recorded her extraordinary testimony, as follows: 'Been down at coal-carrying six weeks; makes 10 to 14 rakes a day; carries full 56 lb of coal in a wooden backit. The work is na guid; it is so very sair. I work with sister Jesse and mother; dinna ken the time we gang; it is gai dark. Get plenty of broth and porridge and run home and get bannock, as we live just by the pit.'[21]

A workmate, Margaret Watson, a sixteen-year-old coal bearer, had also started carrying coals at age six, 'and have never been away from the work, except a few evenings in the summer months'.[22]

A few miles away, Alison Jack, who worked at Loanhead colliery, was an eleven-year-old coal bearer. According to her testimony, she had gone down the pit around the age of eight, working directly for her father:

> he takes me down at two in the morning, and I come up at one and two next afternoon. I go to bed at six at night to be ready for work next morning. I have to bear my burden up four traps, or ladders, before I get to the main road which leads to the pit bottom. My task is four or five tubs; each tub holds 4 cwts. I fill five tubs in 20 journeys. I have had the strap when I did not do my bidding. Am very glad when my task is wrought, as it sore fatigues.[23]

At this point, Franks intervened with a long, detailed account of the girl's working conditions in the steeply inclined bearing pit, as if to make a special point about the abominable toil which was so typical for such workers:

> A brief description of this child's place of work will better illustrate her evidence. She has first to descend a nine-ladder pit to the first rest, even to which a shaft is sunk, to draw up the baskets or tubs of coals filled by the bearers: she then takes her creel (a basket formed to the back, not unlike a cockle-shell, flattened towards the neck, so as to allow lumps of coal to rest on the back of the neck and shoulders), and pursues her journey to the wall-face ... She then lays down her basket into which the coal is rolled, and it is frequently more than one man can do to lift the burden on her back. The tugs or straps are placed over the forehead, and the body bent in a semicircular form, in order to stiffen the arch. Large lumps of coal are then placed on the neck, and she then commences her journey with her burden to the pit bottom, first hanging her lamp to the cloth crossing her head. In this girl's case she has first to travel about 14 fathoms (84 feet) from wall-face to the first ladder, which is 18 feet high: leaving the ladder she proceeds along the main road, probably 3 feet 6 inches to 4 feet 6 inches high, to the second ladder. ... So on to the third and fourth ladders, till she reaches the pit bottom, where she casts her load, varying from 1 cwt. to 1 and a half cwts into the tub. This one journey is designated a rake; the height ascended, and the distance along the roads added together, exceed the height of St Paul's Cathedral; and it not infrequently happens that the tugs break, and the load falls upon those females who are following. However incredible it may appear, yet I have taken evidence of fathers who have ruptured themselves from straining to lift coal on their children's backs.[24]

A seventeen-year-old bearer at Edmonstone colliery, Midlothian, Agnes Moffatt:

> Began working at 10 years of age; works 12 and 14 hours daily; can earn 12 shillings in the fortnight, if work be not stopped from bad air or otherwise.
>
> Father took sister and I down; he gets our wages. I fill five baskets; the weight is more than 22 cwts; it takes me 20 journeys. The work is o'er sair for females; had my shoulder knocked out a short time ago, and laid idle some time.
>
> It is no uncommon for women to lose their burthen, and drop off the ladder down the dyke below … When the tugs which pass over the forehead break, which they frequently do, it is very dangerous to be under with a load.
>
> The lassies hate the work but they canna run away from it.[25]

Jane Kerr, aged twelve, and her sister Agnes, aged fourteen, were bearers in the stair pit at Dryden colliery. The sisters bore coal for their father, but Jane worked six days a week. On those days when her father was not working – usually on Monday and Tuesday – she was directly employed by the owner on a flat day rate, '6d a day for bearing wood for him', in other words, carrying pit props.[26] In those instances, Jane was employed part time as a fremit bearer. This category of female worker is often described as being particularly exploited and abused, and her unenviable situation requires further explanation, as in this critical statement by Robert Bald:

> Besides the wives and daughters of the colliers, there is another class of women attached to some collieries, termed framed bearers, or, more properly, Fremit Bearers, that is, women who are nowise related to those who employ them. These are at the disposal of the oversman below ground, and he appoints them to carry coals for any person he thinks proper, so that they sometimes have a new master every day: this is slavery complete; and when an unrelenting collier takes an ill-natured fit, he oppresses the bearer with such heavy loads of coal, as are enough to break, not only the spirit, but the back of any human being.[27]

We have already seen in Chapter 2 how colliers at Loanhead who had no family helpers for pit work were normally required to employ and house a 'fremit', or 'stranger' bearer. In many cases, colliers were known to grudge

Plate 23.

Dangers of coal-bearing on ladders and stairs.

paying a wage to the fremit bearer, not least as it cut into family earnings, and as non-family bearers were paid slightly more than relatives. In other circumstances, larger collieries often had bearers who, as at Dryden, were paid directly by the colliery itself, and who could be assigned as required, operating as flexible labour and helping to overcome any shortages in the supply of bearers.

Among the few women bearers interviewed by Franks was Isobel Hogg, aged fifty-three, who worked at Penston Colliery in East Lothian. Her testimony revealed important and harrowing details of the many burdens borne from a young age by married women in a collier family and earning unit; as mothers in pregnancy and childbirth, in their domestic roles within the household, and as pit workers enduring foul conditions:

> Been married 37 years; it was the practice to marry early, when the coals were all carried on women's backs, men needed us; from the great sore labour false births are frequent and very dangerous.
>
> I have four daughters married, and all work below till they bear their bairns – one is badly now from working while pregnant, which brought on a miscarriage from which she is not expected to recover.
>
> Collier people suffer much more than others – my guid man died nine years since with bad breath; he lingered some years, and was entirely off work eleven years before he died.
>
> You must just tell the Queen Victoria that we are guid loyal subjects; women-people here don't mind work, but they object to horse-work; and that she would have the blessings of all the Scotch coal-women if she would get them out of the pits, and send them to other labour.[28]

Again, at the end of this testimony, Franks had this to say about his impressive witness: 'Mrs Hogg is one of the most respectable coal-wives in Penston, her rooms are all well furnished, and her house the cleanest I have seen in all East Lothian.'[29]

Jane Peacock Wilson, a forty-year-old coal bearer at Harlaw Muir, supplied much the same heart-rending testimony, including the shocking revelation that the labour of their youngest children underground was considered a necessary help in her prematurely weakened and worn-out state of health:

> I have wrought in the bowels of the earth 33 years. Have been married 23 years, and had nine children; six are alive, three died of

typhus a few years since; have had two dead born; thinks they were so from the oppressive work: a vast number of women have dead children and false births, which are worse, as they are no able to work after the latter.

I have always been obliged to work below till forced to go home to bear the bairn, and so have all the other women. We return as soon as able, never longer than 10 or 12 days; many less, if they are much needed.

It is only horse-work and ruins the women; it crushes their haunches, bends their ankles and makes them old women at 40.

Women so soon get weak that they are forced to take the little ones down to relieve them; even children of six years of age do much to relieve the burthen.[30]

By the early nineteenth century, apart from the collier unions who had called for the removal of all females from pit work, it was the predicament of married women employed as bearers that excited most attention among paternalistic, humanitarian and efficiency-minded owners and managers. In the 1790s, the Ninth Earl of Dundonald had abolished the bearing out of coal at Culross, and had published his views about the need to stop 'the barbarous and ultimately expensive method of converting the colliers' wives and daughters into beasts of burthen'.[31] He had noted the difference in morale, domestic comfort, nurture of their children and more civilised behaviour, brought about by releasing the women bearers for their proper roles in house and home. A few years later, this same emphasis on encouraging domesticity, combined with a passionate concern for the plight of married women bearers and female bearers in general, moved Robert Bald to call publicly for the abolition of this method of carrying hewn coal from the face. In the same style as he had described the toil of the bearers, he set out some of the consequences of retaining this system:

The collier, with his wife and children, having performed their daily task, return home, where no comfort awaits them; their clothes are frequently soaked with water, and covered with mud; their shoes so very bad as scarcely to deserve the name. In this situation they are exposed to all the rigours of winter, the cold frequently freezing their clothes.

On getting home, all is cheerless and devoid of comfort; the fire is generally out, the culinary utensils dirty and unprepared, and the mother naturally first seeks after her infant child, which she nurses even before her pit clothes are thrown off.

From this incessant labour of the wife, the children are sadly neglected, and all those domestic concerns disregarded, which contribute to render the life of the labourer comfortable and happy. It is presumed that it is from this habit of life that infectious diseases make in general greater havoc among the children of colliers than among those of any other class of labourers; so much so, that we have seen the number of deaths in one year exceed the number of births.

How different is the state of matters, where horses are substituted for women, and when the wife of the collier remains at home.

The husband, when he returns from his hard labour with his sons, finds a comfortable house, a blazing fire, and his breakfast ready in an instant, which cheer his heart, and make him forget all the severities of toil; while his wife, by her industry, enables him to procure good clothes and furniture, which constitute the chief riches of this class of the community.[32]

Bald had attempted to abolish bearing at Alloa in the 1820s as the result of a decision to refit the colliery, but his ruling was challenged by the opposition of the colliers and the women workers. Finally, a compromise was agreed to exclude married women from pit work. However, abolition of bearing was followed in many other collieries in the eastern area before 1842, perhaps most notably in Fife, where the evidence suggests that bearing had stopped completely by then. Commissioner Franks did not find any working bearers in his investigation here, and at least two collieries, namely Balbirnie and Dysart, had not employed bearers or any other women workers for several years. However, he interviewed Janet Welch, a former coal bearer at the large Wemyss Colliery, who told him that 'women who worked in the high seam carried coal till masters forbid it two years since' and that she had 'ceased to do so six months ago' at another pit within the colliery complex.[33]

Modernising collieries in Midlothian, for instance, at Whitehill and Arniston, no longer had bearing pits in 1836 and 1840 respectively, and at Dalkeith Colliery the Duke of Buccleuch had ordered that no women be employed. Rosewell and Barleydean collieries stopped employing women and young children in 1837 or 1838. At Newcraighall, near Edinburgh and at Tranent, in East Lothian, although bearing was not completely abolished, and large numbers of girls worked at both collieries, it was apparently the policy to encourage the phasing out of married women bearers. At Sir John Hope's three collieries, including the principal one at Newcraighall with 600 workers, the owner and the oversman had evidently tried to stop the practice of female coal-bearing in 1835. However, as with Bald at Alloa, the initiative was apparently opposed by the male and female workers, who regarded this intervention as an interference with their own work preferences. It is not known whether Hope and the other Lothian masters really intended to do away with bearing and to modernise the workings of the inclined seams in their pits by installing more winding gear, bringing in ponies for haulage, widening the main underground passages and setting down rails. In any case,

Plate 24.
Narrow shaft, winding by one-horse gin, with workers in basket.

there was no technical need to persist with women or child bearers here or anywhere else, as hauling up the coal could be achieved by using the windlass system of ropes and buckets. Franks was eager to prove this point in his own evidence to the enquiry, drawn from his experience of the Welsh coalfield:

> The grievous suffering thus inflicted on so many persons of tender age and of the female sex is perpetuated from the coal-owners continuing to work their mines in modes which have become obsolete in all other districts. ... A little reflection would have prevented a vast deal of unnecessary and painful labour in the working of edge seams in Scotland; for instance, in South Wales (where the stratification is almost vertical), on the sea coast at Britonferry and in the anthracite field in Pembrokeshire, coal-bearing as practised in Scotland is unknown. The coal is transported from the different workings by successive windlasses, or balances, working on inclined planes, which plan obviates the necessity of having recourse to the slavish and degrading employment of female labour at present in practice in the collieries of the East of Scotland.[34]

Yet, as the evidence proves, into the early 1840s, despite some exceptions, the Midlothian and East Lothian pits were still the mainstays of bearing and of conservative methods and practices, as they were also of the cheap labour employment of women and girls. After all, it had always been easier and less expensive to employ female labour and very young small children who were able to crawl through the narrow passages of the older, unimproved mines, rather than investing in horse-powered haulage, additional winding gear and new underground roads. The 1842 Mines Act would force such owners to make the necessary changes in their recruitment practices and, as a consequence, also make accompanying improvements in the efficiency of haulage from the immediate coalface.

Underground Haulage Work

As with coal-bearing on the back, the task of pushing, pulling or dragging quantities of coal from the working face along subterranean roadways to the pit bottom was also achieved by the severest exertion of human labour. To avoid confusion of names and definitions of haulage workers, the word 'drawer' is most commonly and consistently used to denote the worker who took and conveyed the main weight of the filled box or truck to the winding point at the pit bottom. In some accounts, the word 'putter' is often used to describe this task, and may be interchanged with 'drawer', although it can also mean an assistant to the drawer.

Sometimes, this haulage was performed by the hewer or his boy assistant at the face, but was most often done by other young workers, including girls, and by women. Thomas Tancred described the nature of this work as it was performed in the coal mines in the west of Scotland area in 1841–2:

> A numerous race of juvenile workers under ground are the drawers – children, or young persons who drag or push the loaded whirleys along the tram-roads from the place where the coal is worked out to the bottom of the shaft, where it is hoisted up with its load, and the children return with an empty one in its place. Up to the age of 14 or so, two children generally draw together, one pulling by means of leathern hoops through which the arms are passed, having a chain from them hooked to the front of the whirley, and the other, a younger one, pushing behind with both hands.

> It is a general rule that colliers have no right to complain if the roads are kept three feet high, and they are usually a few inches more than this. The amount of labour to which children are subject in drawing is very different, varying in proportion to the load drawn, the distance, and the number of times it is traversed, the inclination of the road, the state of repair in which the rails or tram-ways are kept. The weight of the loaded corf or whirley differs very widely in different mines, but this probably arises from some circumstances in the state of the road rendering the draught lighter or heavier. It is sometimes made of iron, sometimes of wood, but more commonly of wattled hazel-rods secured in a wooden frame, and in all cases running on cast-iron wheels.

> Any one who has seen the children at work can have no hesitation in saying that the physical exertion necessary in drawing is

Plate 25.

Hauling hutch in harness, ropes and chain passing under the pit clothes.

occasionally considerable. The exertion, however, is by no means continuous. The whirley has to be filled, which is in general chiefly done by the collier with a shovel, and by lifting the larger pieces of coal with his hands. The whirley, being loaded and started on the tram-way, runs pretty easily until perchance it gets off the rails at a sudden turn, or where another railway joins in. Then the drawer and his assistant must put their shoulders to the wheel to lift or drag it upon the rails again. After this they can take a little rest. Once more they start and perhaps hear a rattling and see a light in the distance; this is another pair of children trotting along with an empty whirley towards the face of the coal. As there is but one line of rails, the drawer of the loaded carriage halloos to the others to stop, or to turn their whirley off the line. Thus a passage is left for the full one, which proceeds on its way. Now we will suppose they come to a part of the road where there is a slip in the strata, sometimes called 'a trouble'. Here the road rises pretty steeply for a short distance, and now comes the tug of war. The drawer throwing his whole weight upon the chain, and leaning his body so forward that his hands touch the rails, while the putter pushes with might and main behind, with many a puff they urge the load to the top of the ascent. Here they sit awhile till they have recovered their wind, after which they soon see the light dancing about ahead, and hear the hubbub at the pit bottom. Here they have some time to wait their turn, and perhaps eat a part of the food they brought down with them, and pass their joke with the other drawers. This lasts till the bottomer hooks on their whirley to the engine-rope and returns them an empty one, with which they set off at a run back into the mine.

Here, Tancred, in his usual style of understatement when assessing the labours and dangers of this haulage work does not record the testimony of individual witnesses among the workers involved. Nevertheless, he concludes his long, useful passage with the following judgement:

Thus the drawing, though occasionally hard work, admits of frequent periods of rest and refreshment. It affords a varied exercise to the body and limbs, so that I heard no complaints from the children of over-fatigue, or of being oppressed by the workmen for whom they draw, who are usually either their father, elder brother, or some relation; nor do medical men attribute any physical injury to the use of children in drawing. The employment of females in this work, however, or indeed in any capacity below ground, will, I trust, be absolutely forbidden by the legislature.[35]

Wheeled hutches running on cast-iron rails along main roadways leading to and from the working coalface in relatively dry conditions were the norm in the more progressive mines of the west of Scotland by 1840. However, this was still far from being the case elsewhere in the country. Here, although there were well-conducted and modern pits, as at Bo'ness and Carron, Arniston and Dalkeith, the underground working environment in the majority of pits was much worse for many women and child haulage workers. As in coal-bearing, a range of evidence from haulage workers in Stirlingshire, Fife and the Lothians reveals the comparatively primitive, highly noxious, wet, cramped and partially unrailed and undeveloped state of many of the mine workings in which they had to perform manual labour during the early nineteenth century.

In unrailed sections, a sledge, or slype, with a harness pulled by young workers was still in use in some workplaces. The slype was a wood-framed cart, curved and shod with iron at the bottom, holding from two and a quarter to five hundredweight of coal, and adapted to the usually uneven and unpaved floor through which it was dragged. Commissioner Franks described how the lad or lass was

> harnessed over the shoulders and back with a strong leathern girth, which behind is furnished with an iron hook, which attaches itself to a chain, fastened to the coal-cart, and is thus dragged along. The dresses of these girls are made of coarse hempen stuff (sacking) fitting close to the figure; the coverings to their heads are of the same material; little or no flannel is used, and their clothing being of an absorbent nature, frequently gets completely saturated shortly after descending the pit, especially where the roofs are soft.[36]

Where seams were narrow and the roofs low, the young workers dragged the slype on all fours, like horses. The workings in the narrow seams were sometimes 100 to 200 yards from the main roads, so that the worker had to crawl backwards and forwards pulling the small carts in spaces no more than two or three feet high. Franks cited the case of Margaret Hipps, a slype drawer, aged seventeen, who worked at the Stoney Rig Colliery, Polmont, near Falkirk. She worked in several stages, first dragging the loaded sledge to the main road. She then filled the wheeled hutch stationed there, perhaps taking three journeys to do so, before pushing the heavy hutch along the railed stretch, and finished the sequence by crawling back with the empty sledge to the coalface.

> On short shifts I work from eight in the morning till six at night; on long ones until ten at night: occasionally we work all night. Only bread is taken below; and the only rests we have are those we have to wait upon the men while picking the coal.
>
> My employment, after reaching the wall-face, is to fill a bagie, or slype, with 2 and a half to 3 cwt. of coal. I then hook it onto my chain, and drag it through the seam, which is 26 to 28 inches high, till I get to the main road – a good distance, probably 200 to 400 yards. The pavement I drag over is wet, and I am obliged at all times to crawl on hands and feet with my bagie hung to the chain and ropes.
>
> I turn the contents of the bagies into the carts till they are filled, and then run them upon the iron rails to the shaft, a distance of 400 to 500 yards.
>
> It is sad, sweating and sore fatiguing work, and frequently maims the women. My left hand is short of a finger, which laid me idle four months.[37]

Janet Moffatt, a twelve-year-old drawer at Newcraighall Colliery, worked in similar, miserable conditions:

> Works from six morning till six night: alternate weeks works in the night shift. Descends at six at night, and returns five or six in morning, as the coals are drawn whiles later. I pull the wagons, of 4 to 5 cwt., from the men's rooms to the horse-road. We are worse off than the horses, as they draw on iron rails and we on flat floors.

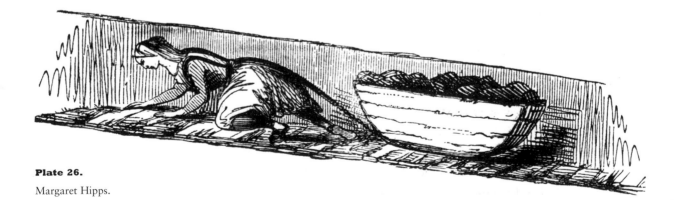

Plate 26.
Margaret Hipps.

We have no meals below. Some of us get pieces of bread when we can save it from the rats, who are so ravenous that they eat the corks out of our oil-flasks.

Draw the carts through the narrow seams. The roads are 24 to 30 inches high; draw in harness, which passes over my shoulders and back; the cart is fashioned to my chain.

The place of work is very wet and covers my shoe tops.

I work on mother's account with sister, as father was killed in the pit five years since. There are often accidents below: a woman was killed 3 months since by one of the pit waggons.[38]

After providing these testimonies, an outraged Commissioner Franks was again moved to write the following indictment:

It is almost incredible to believe that human beings can submit to such employment, crawling on hands and knees, harnessed like horses, over soft slushy floors more difficult than dragging the same weights through our lowest common sewers, and more difficult in consequence of the inclination, which is frequently one in three to one in six … The dangers and difficulties of dragging on [such] roads may be more easily conceived than explained; and the state that females are in after pulling like horses through these holes – their perspiration, their exhaustion, and very frequently even their tears, it is painful in the extreme to witness.[39]

Elizabeth Gibb, a twelve-year-old drawer, and one of many girls employed at Fordell Colliery, in Fife, exemplified the tribulations of this class of child workers:

draws the corves with chains; not harnessed; holds the chain in both hands and draws forward, like the horses; has done so four years. Makes 60 to 70 races (journeys) every day. The work is very hard. Once split my finger and was idle one month; and afterward broke my leg by the overload of a waggon, and was idle three months.[40]

Some of the girl and women drawers were former coal-bearers who either by choice or necessity had switched over to cart and wagon haulage in places where bearing had been abolished. Margaret Galloway, a thirteen-year-old worker at East Bryants Colliery, Midlothian, was both drawer and occasional coal-bearer. 'The work fatigues me much and often crushes me. The roads are very wet; at parts the water passing the ankles, and frequently higher, so that our lower clothes are quite wet.'[41]

From a different age group, Jane Johnson was then a 29-year-old married woman with four children, who worked at Preston Hall colliery, in Midlothian. She had worked since the age of seven, firstly at Sheriffhall, as a bearer and could carry two hundredweight when she was a fit fifteen-year-old. Since the change to drawing carts, 'I now feel the weakness upon me from the strains. Many women lose their strength early from overwork, and get injured in their backs and legs; was crushed by a stone some time since, and forced to lose one of my fingers.'[42]

Like the fremit bearers, some teenage girls, still physically fit, but already feeling the adverse effects of heavy haulage labour, did not work directly for the male members of the collier family unit. Instead, they were contracted by the colliery managers, or subcontracted by hewers, with the agreement of the oversman, to do this work on a day- or piece-rate basis. This situation prevailed at the large Wemyss collieries, as the colliery agent explained in April 1841: 'The coals are putted by females, and the practice here is to contract with a certain number of responsible hands for periods of three and six months, leaving these contractors to engage their assistants.'[43]

However, being exploited by such contractors created its own additional pressures on the haulage assistants who did the heaviest work. Elizabeth Lister, a fifteen-year-old drawer, complained:

> when I wrought on a day's wages for master, was not so hard worked; the work is more sair, as the men drive us more, for they do the work cheap. Many girls have left, not liking to be driven, and gone into the fields.
>
> Works from six in the morning till six at night; has to make 14 races before porridge time; the distance is 300 fathoms from incline to pit bottom; and 14 and 15 races between porridge and the time we take our pieces of bread; 14, 15, and 16 races afterwards; we get 15

Plate 27.

Drawing in Mid and East Lothian.

[96]

pence a day but only employed nine sometimes ten days in the fortnight.[44]

Isabel Hugh, a nineteen-year-old drawer, also worked at Wemyss on a direct contract arrangement, with a female co-worker. It is interesting to see that they also had an arrangement to do additional maintenance work:

> Began to work when 13 years old, below ground; has wrought in the fields; likes the work well enough; it is guid sair sweating work. Janet Adamson and I contract for putting on our own account; the road is 100 fathoms in length, and we run the races singly; we frequently run 50 races between us; we get 14p per score and 1/- per week each extra for clearing pit bottom and working the pump; seldom work less than 12 to 14 hours.[45]

Some instances of women and girls doing other oncost work below ground were noted by Franks. One such was Phillis Flockhart, a twelve-year-old night maintenance worker and occasional bearer at Edmonstone Colliery, Midlothian, who related her circumstances:

> I work with the redesmen [road-clearers] who go down at night to clean the roads and make the walls; I bear the bits of stone for the wall building, to keep up the roof. I have wrought at the work 12 months; have been at coal work whiles; was bringing coal along the pit some months since, when I got the flesh torn off my leg, and was idle seven weeks. I work in No. 14 pit on aunt's account. I am a natural child; mother left me when three years of age, and aunt has kept me ever since. … she being now too old to labour.[46]

In the new ironstone pits of West Lothian and at Shotts and Whiterigg, near Airdrie, in the eastern parts of Lanarkshire, small numbers of women and

Plate 28.

Drawing in Fife and Clackmannan.

girls were employed in haulage work. Hutch loads of ironstone were heavier than coal, and at Whiterigg, women hauled huge loads of up to eleven hundredweight (half a ton). Very young girls were usually excluded from that work, although Franks found twelve-year-old and teenage girls at Crofthead Mine, near Whitburn. Mary Brown, a thirteen-year-old drawer and former coal-bearer, whose family had migrated from Midlothian, supplied a full profile of her compulsory working life to date:

> Wrought in Crofthead ironstone mines six months. The work is very heavy and sore fatiguing, as I have to shove 15 to 20 hutches every day, and the distance is far away from the shaft. The weight of ironstone varies in hutches, sometimes 5 cwt., at others 8 cwt. The roads are all well railed; would prefer daylight work better.
>
> Has worked near five years in mines; was last at Sir George Suttie's, at Prestongrange, as a bearer of coals in that part where machinery was not employed, it being too steep for putting. There are many children work below; none of them like it, nor would they go down but their fathers or elder brothers force them.[47]

Young mineworkers were forced into doing other jobs within the pits, in circumstances where they would rarely have volunteered. They were not usually paid directly, the money going to their collier parent or guardian. Boys were employed as pony or horse drivers in underground haulage and at the pit-head. Robert Thomson was an eleven-year-old horse driver at Newcraighall in 1841: 'Drives a pony in the Tunnel Mine; works 12 and 14 hours; has done 18 months. Would like it fine if the time would allow me to see the daylight. I work to father. The pit is very wet and sair drappie.'[48]

Underground water pumpers and trappers at the ventilation doors were also generally very young and poorly paid, working some of the longest hours in miserable conditions, and the pumpers often being immersed in filthy water. It was mainly boys who did this work, and their only hope was to graduate sooner or later to different and better jobs within the mine. Alexander Gray, aged ten, was a pump boy at Newcraighall, working in the deepest part of the pit to clear rising water to the level of the engine-pump:

> I pump out the water in the under bottom of the pit, to keep the men's rooms dry. I am obliged to pump fast or the water would cover me. I had to run away a few weeks ago, as the water came up so fast that I could no pump at all, and the men were obliged to gang. The water frequently covers my legs, and those of the men when they sit

to pick. I have been two years at the pump. I work every day, whether men work or not. Am paid 10d a day: no holidays but Sabbath. I go down at three, sometimes five in the morning, and come up at six and seven at night. I know that I work 12 and 14 hours, as I can tell by the clock.[49]

James McKinlay, aged nine, also worked at Newcraighall as a pump boy: 'I gang below with my two sisters at three in the morning. We take bits of bread; we get nothing else until we return at three and four in the day. We work all night week about. Father gets 10d a day for my work.'[50]

Commissioner Franks reckoned that the job of the trapper was one 'of the most monotonous and deadening to all the mental and physical powers of a young child. The trapper has to sit, often exposed to damp, completely in the dark and in silence, from the time the coal begins to be brought forward by the drawers till the last whirley has passed, cheered only by the occasional gleam of a lamp from a passing whirley, or a few words with the drawers.'[51]

Thomas Duncan, aged eleven, worked as a trapper at the Marquis of Lothian's East Bryants Colliery:

I open the air doors for the putters; do so from six in the morning to six at night. Mother calls me up at five in the morning and gives me a piece of oatcake, which is all I get till I return; sometimes I eat it as I gang. There is plenty of water in the pit; the part I am in comes up to my knees. Mother has always worked below; but father has run away these five years. I get 3 shillings a week, and take it home to mother.[52]

The final sample of evidence of the employment of girls in pit work from the vantage point of 1841 is taken from William Baird's Gartsherrie colliery, Coatbridge, apparently the most western location for known female pit employment at that late date. Evidently, following a recent successful battle to uproot collier trade unionism from their pits, the owners had resorted to the re-employment of women and children. Thus, Janet Snedden, aged nine, a trapper in the No. 1 pit, 'comes down with Janet Ritchie, a single woman who hooks on and off the corves [hutches] on the chain for drawing coal up the pit; comes down a quarter before 6 and goes up again about 4pm', the long working day of the girl and woman coinciding with the starting and finishing time of the main colliery winding engine.[53]

Clearly, before 1842, there was no typical colliery in terms of size, equipment, efficiency, employment and working practices within the existing

Scottish coalfield area. However, the bigger, the more modern and the more progressive the colliery, the greater was the division of labour within the workforce below and above ground. Moreover, in those collieries the workforce was already all male, or nearly so, and especially below ground. At the pit-head, there were various craft trades and ancillary occupations among surface workers in a large and well-equipped colliery such as Govan by the early part of the nineteenth century. They can only be listed and mentioned here to complete our colliery profile.

Enginemen or brakesmen operated the pumps and winding engines, and the cleeksman saw to the arriving and descending tubs, while a pit-headman weighed the loaded tubs and noted the tally due to each hewer for his work. Banksmen unloaded the tubs, and either emptied them into waiting wagons and carts, or tipped the load on to the stockpile. Some elementary coal sorting was done before that stage, especially to separate out the small coal and dross. Unlike later, when women were usually employed at screening and sorting, men and boys did those jobs.

Surfacemen with responsibilities for maintenance of underground workings, roadways, rails and shafts included brickmakers, masons and blacksmiths, carpenters and joiners for making and mending tubs and sawyers for timber and props. Workshop stores provided the collier's tools and equipment, such as picks, powder, candles and oil, all of which were subject to payment (offtakes) from the worker. In the bigger collieries, the conspicuous scene at the surface would include many horse and pony keepers in charge of large numbers of these essential animals for haulage and carting tasks. Underground pony driving and stabling were also regular features of colliery working-life by this time, involving men and boys. It would remain so into the twentieth century as a reminder that human and horse power could not be completely supplanted by the machine in the enterprise of winning and transporting coal.

A Black Country: Capital and Labour in an Expanding Mining Frontier, c.1830–1880

The Mining Frontier: An Overview

'Oor Location'

A hunner funnels bleezin', reekin',
Coal an' ironstane, charrin', smeekin';
Navvies, miners, keepers, fillers,
Puddlers, rollers, iron millers;
Reestit, reekit, raggit laddies,
Firemen, enginemen, an' Paddies;
Boatmen, banksmen, rough an' rattlin',
'Bout the wecht wi' colliers battlin',
Sweatin', swearin', fechtin', drinkin',
Change-house bells an' gill stoups clinkin'.[1]

The self-taught local poet's dialect voice is an appropriate topical starting point for this chapter. Written in the late 1850s, Janet Hamilton's evocative verse conveys a strong impression of the clamour and turbulence of the industrial scene as directly witnessed in the coal-mining and 'iron burgh' of Coatbridge. It also captures a sense of the tensions and conflicts within an industrial community, and particularly in that social melting pot created by the influx of thousands of incoming workers (including many Irishmen) who crowded into the Monklands district of north-east Lanarkshire. They had flocked in to work in the mines and iron works of Scotland's first extensive 'black country' district.

Coatbridge, in the middle of Old Monkland parish, was already the nerve centre of the new hot-blast iron industry in the 1830s. Its six iron works and fifty blast furnaces produced top-class foundry iron, based on the mining and smelting of abundant rich resources of splint coal and untapped seams of blackband ironstone. In a hectic rush to buy and rent mineral leases, the ironmasters began the ruthless exploitation of local deposits from an area extending over 4,000 acres in the parishes of Old and New

Monkland, which included the growing town of Airdrie and its sprouting, surrounding pit villages.

The Old Monkland furnaces were consuming over half a million tons of coal annually by the end of the 1830s. By 1840, the ironmasters led the additional expansion of coal and ironstone mining in North Lanarkshire into the nearby parishes of Bothwell, Shotts and Cambusnethan, forming the crescent of the northern section of the great central coalfield. It supplied the vital mineral fuel for the Lanarkshire hot-blast furnaces and the many new wrought-iron works – with their 'hunner funnels' – also concentrated in Coatbridge by mid-century.

The pig-iron and wrought-iron works created an enormous demand for coal and ore and a major social consequence was the dramatic and continuing increase in the working population, most notably in mining. The working class of Monklands had doubled to 40,000 between 1831 and 1841. The largest section of wage earners was engaged in coal and ironstone mining, followed by ironworkers and general labourers. To demonstrate the scale and impact of mining development, in August 1842, when Lanarkshire coal and ironstone miners staged a massive general strike against wage cuts, over 140 pits in the Airdrie and Coatbridge district alone were directly affected. An estimated 8,000 mineworkers from a total of 10,000 took part in the strike in Lanarkshire. The census figure for 1841 shows that Lanarkshire miners then formed half the total of mineworkers within the

Plate 29.

Gartsherrie blast furnaces 'old side', built in 1830s. (*North Lanarkshire Council*)

whole Scottish coalfield area, and continued to this extent until the 1870s when they numbered over 25,000.[2] In the early 1840s, the rash of over 100 pits in a relatively small locality indicates the feverish amount of prospecting that was taking place in mining, involving a great deal of speculation aimed at making a quick fortune. Short-lived ventures were typical, and shallow, single pits were worked and abandoned, existing alongside the extensive, persistent and more productive enterprise of the few, large coal and iron companies. In this insecure mining frontier, with a high labour turnover, many mineworkers had to be mobile and migratory in their pursuit of regular work and tolerable conditions. However, despite the uncertain conditions facing mineworkers, Old Monkland parish was already becoming the most heavily industrialised parish in mid-Victorian Scotland. Here, the scale and concentration of mining, iron manufacture and associated metal engineering was to reach its greatest extent in the boom period of the early 1870s, encompassing over twenty ironworks of various descriptions and forty collieries and ironstone mines employing more than 3,000 pitmen.

In the 1830s and 1840s, the nature of this raw, mushroom growth, with its industrial frontier within North Lanarkshire, particularly in the sprawling working-class suburbs of south and east Glasgow, and the adjacent parishes of Old and New Monkland, provoked alarm and fear among respectable society and authority. For them, this frontier area was the shock centre of a new world inhabited by a dangerous mass of unruly workers who appeared to threaten the very existence of civilised society. In his report to the 1842 Commission, Thomas Tancred referred to the Monklands industrial frontier. He expressed concern about the large motley population of incomers, hastily thrown together, whose overcrowding was already putting an enormous strain on local resources and generating massive social problems:

> This vast and sudden accession of population, consisting for the most part of irregular and dissolute characters from all parts – from Wales, England, Scotland and Ireland. … At Coatbridge, where a large portion of this population has been located within the last ten years, no church or clergyman has been supplied them till very recently, when a church was erected, chiefly at the expense of one of the numerous employers of labour in the district. There is also a relief church, provided by voluntary contributions. These efforts come, of course, as must always be the case so long as things of this importance are left as now to accident and chance, too late. In the meanwhile, a population has been growing up immersed more deeply than any I have met with in the

most disgusting habits of debauchery. I feel that my powers of description are wholly inadequate to convey the feeling inspired by a visit to these localities. Everything that meets the eye or ear tells of slavish labour united to brutal intemperance. This population consists almost exclusively of colliers and iron workers, with no gentry or middle class beyond a few managers of works and their clerks. I visited many of the houses attached to some of the works, and usually found them in a most neglected state, bespeaking an absence of all domestic comfort or attention to social duties. The garden ground usually lay a mere waste, unenclosed, and not a spade put into it; the children, in rags and filth, were allowed to corrupt each other, exempt from all the restraints of school or of domestic control. This domestic discomfort seemed attributable, amongst other causes, to the crowded state of the habitations, which, from the want of buildings to contain the rapidly increasing population, were filled with lodgers. I was assured that some houses, with a family and only two rooms, took in as many as 14 single men as lodgers. It is needless to observe how impossible it must be for a woman to preserve decency, cleanliness, or comfort, under such circumstances. An infatuated love of money, for no purpose but to administer to a degrading passion for ardent spirits, seems the all pervading motive for action in this quarter.[3]

Seymour Tremenheere, the mining commissioner who officially inspected the Monklands on a regular basis for several years after 1842, continued in a similar vein, giving vent to his class prejudice, as if this emerging working population was an irresponsible subspecies:

The vast collection of people congregated in these two parishes of Old and new Monkland, to the amount of 40,000, is composed with comparatively few exceptions, of the class of colliers, miners, and men working about the furnaces. Whatever restraint therefore may naturally be supposed to be exercised by the supervision, authority and examples of the classes of a higher grade, is but little felt there. The low habits of the uneducated or demoralised receive little check or rebuke from a superior presence. These outbursts of dissipation are greatly aggravated by two circumstances – the want of an efficient police and the unlimited facilities for obtaining ardent spirits.[4]

On the latter points raised by Tremenheere, there was certainly cause for major concern. Whatever the underlying reasons for the alleged lawless and degraded behaviour among mineworkers, unreasonable conduct was nevertheless exacerbated by excessive consumption and ready availability of cheap alcohol. It was also common practice for workers to be paid in public houses. In the early 1840s, the mining district of the Monklands was infested with an estimated total of 330 public houses of various sorts, or about one for every twenty adult males, and many remained open at all hours. On Saturday pay nights, we are informed that Airdrie was invaded by hordes of miners from the surrounding pit villages, intent on a drinking spree that would last through to Sunday; and that the burgh's five-man police force had to face 'scenes of uncontrolled license which there are no means of either preventing or punishing'.[5]

Hostile comment was also forthcoming from the region's leading law officer, Archibald Alison, sheriff of the county of Lanark, who was based in Glasgow. A cultured middle-class Tory, he was appalled by the rough and volatile behaviour of the mass of mineworkers. As the 'man on the spot' responsible for defence of public order in times of emergencies, such as the outbreaks of widespread discontent during the miners' strikes of 1837 and 1842, Alison claimed some understanding of the wild frontier society he had to encounter. In over twenty years as sheriff, Alison would meet his greatest challenges in the Monklands, an episodic storm centre of unrest among the miners. He had analysed the dramatic and unprecedented population changes of recent years, noting that the parishes of Monklands and Bothwell had experienced 'an American rate of increase' comparable to similar frontier communities that had sprung up so suddenly 'on the banks of the Ohio or the Mississippi'. As in these communities, his Lanarkshire domain had few lawmen to police over 200,000 people, including the working class of several growing towns and many scattered, newly formed rural mining settlements. According to his perception, the mining settlements contained an underclass of savage, ignorant, heathen, drunken, work-shy and debased males who, when acting collectively, as in disorderly industrial action, were a menace to private property and public order and had to be put down by military force, if necessary. Alison was also scathing in his contempt for the large employers and propertied interests in the county for their failure to agree to pay for sufficient policing, to enforce licensing laws and to create an adequate infrastructure of schools, churches and housing provision as civilising influences to 'stave off the tide of irreligion and immorality'.[6]

Of Coatbridge, the ironmaster Robert Baird, fearing for his safety and his property, told the mines commissioner, 'there is not a worse place out of hell'.[7] Considering the conditions and grievances endured by the majority of mineworkers and their families in this neighbourhood, they would have had no trouble in agreeing with the ironmaster, albeit from a very different perspective, that their own experience of insecurity was living hell itself.

From the 1830s until the 1870s the big ironmasters dominated mining in the Monklands, Lanarkshire and the west of Scotland. In the West, they had taken over from local landowners as the principal owners and operators of mining enterprises, concentrating their effort on extraction of furnace coal and quality ore for industrial needs. This ongoing quest for mineral fuel initially prompted the Monklands ironmasters to extend and expand their sphere of influence further into Lanarkshire, particularly into the Motherwell and Wishaw districts. In the wake of the Wishaw and Coltness Railway, which they had largely financed, the mining and industrial frontier was opened up in the 1840s from the Monklands out to the Coltness Iron Works, at Newmains. By the 1860s, this railway was only one of many in a network of lines stretching across central Scotland from Ayrshire to the Lothians, accessing and linking all the established and new mining areas alike, and drawing in work and settlement to many hitherto isolated areas. In the 1860s and early 1870s, the ironmasters also secured mineral leases across the Clyde

Plate 30.

Blast furnace plants, nineteenth century.

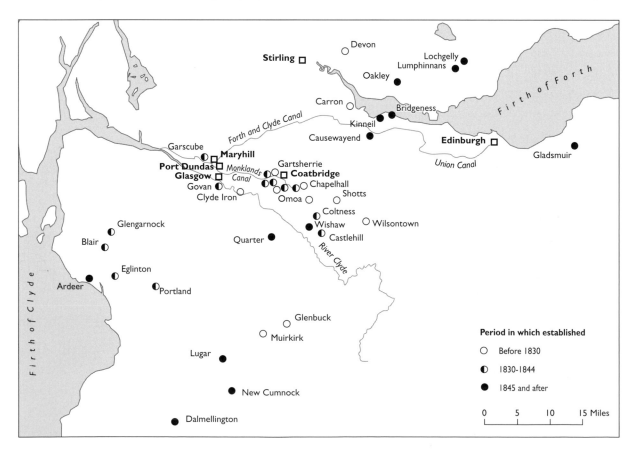

into the Hamilton, Blantyre, Bothwell and Cambuslang districts. Here, a few of the deepest pits in the west of Scotland were sunk to exploit thick seams of furnace coal. For example, Dixon's Blantyre colliery (the location, in 1877, of Scotland's worst mining disaster) extracted coal from a depth of 780 feet.

The industrial landscape of Ayrshire was also being transformed from the 1840s, as the Lanarkshire ironmasters expanded their empire into the county. They established or took over several iron works, particularly in the north around Irvine, Kilwinning and Glengarnock; in the east around Cumnock; and in the south, at Dalmellington. William Baird and Company and Merry and Cunningham were the most prominent and powerful incomers. They proceeded to exploit extensive resources of blackband ironstone and splint coal for their pig-iron works, and employed and controlled thousands of mineworkers in new company settlements around and beyond the iron works. By 1870, there was hardly a coal-bearing Ayrshire parish whose minerals were not being worked for the benefit of an iron company; although sale coal was also important, much of it destined for export.

Lanarkshire and, to a lesser extent, Ayrshire were the principal locations of the expanding mining frontier, and between them the source of the bulk of total coal and ore output in Scotland between the 1830s and 1870s. Other counties were also penetrated by the Lanarkshire iron companies in the search for mineral fuel. William Baird and Company had collieries in all the mining counties of the west, except Renfrewshire, where Merry and Cunningham were active mining operators. To the north, the Monklands influence was felt around Kilsyth and Denny, where ironstone reserves were exploited. In West Lothian, from the late 1830s, several Lanarkshire iron companies were heavily involved in opening up coal and ironstone mining immediately across the county border, and around West Calder, Bathgate and Armadale. Bathgate parish experienced a population increase of 10,000 between 1841 and 1861, the largest occupational group being incoming coal and ironstone workers employed by the iron companies. Further east, by the 1860s and 1870s, Lanarkshire companies had taken out ironstone leases in Midlothian, notably around Penicuik. However, their influence as employers of industrial labour was insignificant in this county, as the resident landed gentry dominated the mining sector.

Beyond the Monklands connection, iron manufacture and associated mining was pioneered in the Fife coalfield, especially around Lochgelly, although this enterprise was on a relatively small scale, and was dwarfed by the traditional emphasis placed on the export of coal.

By the late 1850s, an entirely new mining venture had appeared on the industrial frontier, centred mainly in the parishes of the Almond valley in West Lothian. This was the mining of shale, a rock of compressed mud that, after being roasted in furnaces, produced a crude oil. Distilled and refined, it produced paraffin and lubricating oil for paint and other uses. At first, pioneered by James 'Paraffin' Young in the early 1850s, local cannel coal was mined and processed for oil products. However, in the 1860s, this oil-bearing gas coal was replaced by the greater reserves of shale, and a mania of prospecting led to the expansion of mining, formation of companies in the local oil industry and an increased pool of workers, including incomers from western counties and Ireland. For instance, under the combined influence of coal, ironstone and shale mining, West Calder became a booming frontier parish, its population swelling from less than 2,000 in 1861 to nearly 8,000 in 1871. However, although shale and its associated oil plants had begun to form an important feature of the West Lothian economy in the 1860s and 1870s, the scale of shale mining was still comparatively small, occupying less than 1,000 mineworkers.[8]

Although the lead-mining frontier expanded slightly during this period of the nineteenth century, almost all of the ore raised in Scotland (97 per cent) came from the Leadhills and Wanlockhead workings. Mining enterprise at Islay, Strontian and Tyndrum continued to be irregular and unprofitable, and prospecting at almost twenty other sites proved unsuccessful. There were two exceptions, both in Galloway, at Carsphairn, opened in 1839, and at Cairnsmore in 1845, where small-scale production contributed to an overall total of only 2,000 tons of ore annually by mid-century, rising to around 4,000 tons in the 1860s and 1870s. Around eighty miners from Leadhills had migrated to Carsphairn at the start of that venture, attracted by better prospects, and a lead mining community was formed there. In Scotland, the number of lead miners increased marginally throughout the period, over 500 being concentrated at Leadhills and Wanlockhead. As in shale mining, this still small sector had less than 1,000 mineworkers.[9]

In Scotland, coal and ironstone mining and its frontier of operation had extended considerably between the 1830s and 1870s. Total output had increased fivefold. Around 50,000 mineworkers produced 17 million tons of coal from over 400 collieries in the peak boom year of 1873. This compared to a yearly average of 3.2 million tons in the early 1830s, produced by a workforce of 15,000. In the mining sector overall, the west of Scotland was firmly in the lead, headed by Lanarkshire, with ownership and control in the hands of the large iron companies.

The Mineworkers, 1840s–1870s:
A Changing Profile

The Mines Act of 1842 prohibited women and girls from pit labour and set a minimum age of ten for the employment of boys. In Scotland, an estimated 2,400 women and girls were employed underground when the Act was passed in August 1842. The Act came into force on 1 March 1843, by which time owners, managers and workers were expected to make the necessary adjustments to comply with the law. The majority of masters and workers complied with the employment ban, but there were also notable cases of opposition and evasion, particularly in Midlothian, the Falkirk and Clackmannan areas, and a few instances in Lanarkshire (Shotts and Coatbridge), and around Dunfermline. Although disclaiming any direct responsibility, Sir George Clerk, mine-owner in Midlothian and a minister in the Tory Government, was implicated in a public scandal, when it was revealed that females were still working underground in his mines in 1843, after the legal deadline. The Duke of Hamilton was implicated in the same way, at his Redding colliery, in Falkirk. The two grandees had to give assurances to the Mines Commissioner that the law would be observed.[10] Even then, although large numbers were not involved, until at least 1848, illegal female employment remained a problem at Redding and at one or more of the Carron Company collieries. Eventually, the practice ended at Carron, when the managers ordered a stoppage of work each time a female was discovered at work in the pit.[11] By mid-1844, Tremenheere reported that systematic contravention of the employment ban had ceased, and was monitoring a situation where only 200 or so females were still employed illegally at several places. This figure is a likely underestimate, due to difficulties in reporting and detection. He was the sole commissioner, visited only once a year, announced his visits beforehand and did not conduct any inspections underground. Final enforcement came only after the appointment of mines' inspectors in 1850.

At the start, hundreds of women had continued to evade the law and sought to continue pit work because they needed the money for themselves or their families. The employment ban could not have come at a worse time, as mining and industry were in the middle of a slump in 1842, wages had been reduced and able-bodied men and women who were put out of work were not eligible for parish relief in Scotland. The predicament of unmarried pit women and girls was one of severe deprivation, as they had worked as beasts of burden underground for most of their lives. They were untrained and considered unsuitable for other waged work such as domestic service,

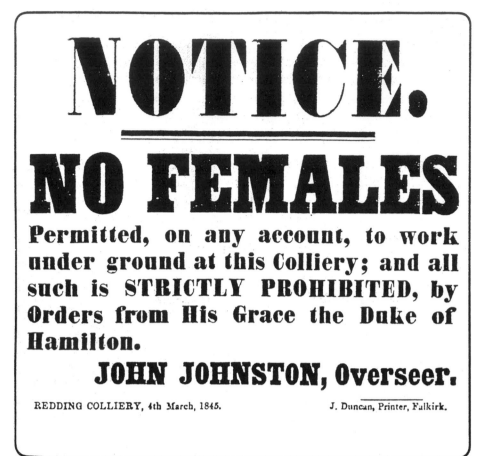

NOTICE.

NO FEMALES

Permitted, on any account, to work under ground at this Colliery; and all such is STRICTLY PROHIBITED, by Orders from His Grace the Duke of Hamilton.

JOHN JOHNSTON, Overseer.

REDDING COLLIERY, 4th March, 1845.   J. Duncan, Printer, Falkirk.

and in many parts of the coalfields there was little or no available alternative waged work.

Desperate to continue, regardless of regulations, female workers had risked entry to the mine by stealth, and several had disguised themselves as men. In many such instances, colliers, contractors and managers, acting out of sympathy or convenience, had been known to turn a blind eye to their continuing employment.

Most of the unmarried women and girls did not find waged work in the short term. In Clackmannan, some were taken on at the woollen mills; in Fife, some found work at the brick kilns and farms. Others elsewhere were placed in domestic service, often after their former employers had exercised some responsibility for their plight. For example, in Midlothian, colliery-sponsored training in sewing and domestic skills was provided at industrial schools to fit the young women for future roles in the home and service.[12]

Overall, as yet, few unskilled jobs for the excluded women were created or found at the pit-head, where men and boys were generally employed. However, from 1842, at some Scottish collieries, women and girls began to

Plate 31.

Colliery notice, 1845.

replace male surface workers in tasks such as basic coal-sorting, moving and tipping hutches.

A historic piece of protective legislation, the Mines Act of 1842 was without doubt welcomed by most women, girls and young boy workers who had hoped and waited for the day when they could somehow be released from such miserable toil. In the case of married women with young family, the legislators of the Mines Act intended such women to drop out of the labour market completely, and take up their 'natural sphere' as full-time homemakers. The evidence suggests that this did occur, any serious financial loss arising from their redundancy from pit work being questionable. In a typical example, a former colliery woman from Pencaitland, who had four children, perceived the benefits of the change as follows:

> While working in the pit I was worth to my husband 7s. a week, out
> of which we had to pay 2s.6d. to a woman for looking after the
> younger bairns. I used to take them to her house at 4 o'clock in the
> morning, out of their own beds, to put them into hers. Then there
> was 1s. a week for washing; besides there was mending to pay for,
> and other things. The house was not guided. Then when I came home
> in the evening, everything was to do after the day's labour, and I was
> so tired I had no heart for it; no fire lit, nothing cooked, no water
> fetched, the house dirty, and nothing comfortable for my husband. It
> is all far better now, and I wouldna' gang down again.[13]

The Mines Commissioner also reported a mixed reception among male colliers to the ban on female pit employment at their workplaces in the east of Scotland. In some places, resentment arose when managers took opportunities to discharge men from unskilled surface work and replace them with pit girls at lower wages. Yet, concerning the enforced removal of females from underground work, he reckoned that, 'a large proportion of the men approved of the alteration', mainly on the grounds that the deployment of men, boys and ponies in haulage operations resulted in more regular loads being taken to the surface, thus increasing their earnings potential.[14] However, in east and central Scotland, many male colliers must also have been converted or impressed by the recent experience of an improved home life, occasioned by the additional full-time presence there of their wife or a daughter, and had began to accept this change as permanent.

From the 1830s until the 1870s, a threefold increase in mineworkers in Scotland indicated the rising demand for labour, which was met by internal

and external recruitment. By the 1840s, when the workforce underground eventually became all male, it was already evident that the majority of recruits into the most rapidly expanding coalfields of the west no longer came from the ranks of the traditional, native stock of home-bred Scottish colliers. However, from that particular source, the son working with the father, or a boy with other male colliers within the family unit, did provide a regular flow of intake. From 1842 until the early 1870s, boys from age ten to late teens were increasingly in demand for several pit tasks, according to their age and strength, and a collier father was expected to take at least his firstborn son down the pit to work alongside him. As more attention was being paid to improved ventilation, the youngest boys continued to find work as trappers, until the opening and shutting of doors was made a self-acting process before the end of the century. This was arguably the lowest-grade task in pit work after mid-century, and the sons of colliers usually managed to avoid being placed there. Orphans or boys from poor mining and non-mining backgrounds alike could find themselves in this work at an early age, their pittance of a wage being a necessary part of the family income. To support his mother, Keir Hardie started as a ten-year-old trapper at the Moss pit, near Newarthill, in 1867. His progress up through the hierarchy of underground jobs followed the standard pattern, becoming a pony driver two years later and, after his strength increased, rising to become an assistant hewer at the coalface.[15]

According to the Mines Commissioner in the mid-1840s, the law regarding the minimum employment age of boys in pit work was being observed in Scotland. In any case, it was becoming recognised that there was no longer any need or justification, on grounds of productivity and safety, to take on boys at this tender age, although the young trapper's job remained an unfortunate exception. Among older and fitter pit lads, especially within the 13–16 age range, heavy haulage tasks, hewing, filling and otherwise assisting at the immediate coalface were the most obvious and numerous roles in this expanding workforce of young men. It was not until the Mines Act of 1872 that the minimum age for pit work was raised to twelve years, and compulsory half-time schooling was introduced for pit boys over that age. However, before then, in a few rare cases, colliery proprietors had already ordered that boys would not be employed before age twelve. For instance, at the progressive Newbattle collieries, as early as 1844, boys began work at twelve years, but only on condition that they could read and write to a basic level, after attending the colliery school there or at nearby Whitehill.[16]

Raw recruits to mining in the 1830s and 1840s came from several external sources. Reluctant to turn to mining but desperate for work, some

Numbers of Miners in West of Scotland 1841–1871

	1841		1851		1861		1871	
Ayrshire	2,526		8,004		11,737		12,600	
	Coal	2,320	Coal	6,061	Coal	7,716	Coal	8,207
	Iron	172	Iron	1,943	Iron	3,706	Iron	3,041
	Unspec	34	Unspec		Unspec	315	Unspec	1,352
Lanarkshire	9,634		19,225		20,549		26,456	
	Coal	7,391	Coal	15,580	Coal	15,101	Coal	22,663
	Iron	1,702	Iron	3,645	Iron	3,827	Iron	2,594
	Unspec	541	Unspec		Unspec	1,215	Unspec	1,199
					Owners / Managers / Clerks	406		
Renfrewshire	925		965		1,799		1,593	
	Coal	910	Coal	790	Coal	610	Coal	433
	Iron	6	Iron	175	Iron	1,065	Iron	733
	Unspec	9	Unspec		Unspec	124	Unspec	427
Dunbartonshire	251		573		620		1,414	
	Coal	205	Coal	549	Coal	504	Coal	675
	Iron	31	Iron	24	Iron	74	Iron	660
	Unspec	15	Unspec		Unspec	42	Unspec	79
Stirlingshire	1,232		2,354		3,551		4,582	
	Coal	1,075	Coal	2,233	Coal	2,703	Coal	3,783
	Iron	113	Iron	121	Iron	650	Iron	786
	Unspec	44	Unspec		Unspec	135	Unspec	13
					Owners / Managers / Clerks	63		
Totals for region	14,568		31,121		38,256		46,645	
	Coal	11,901	Coal	25,213	Coal	26,634	Coal	35,761
	Iron	2,024	Iron	5,908	Iron	9,322	Iron	7,814
	Unspec	643	Unspec		Unspec	1,831	Unspec	3,070
					Owners / Managers / Clerks	469		

Plate 32. Numbers of miners in west of Scotland, 1841–71. (*G Wilson,* Alexander McDonald)

came from the ranks of displaced and depressed handloom weavers, whose livelihoods had been destroyed by over-production, low wages and the advent of the steam-powered loom. Irishmen were included within this category of destitute weavers, having settled earlier in various textile communities in Ayrshire, Renfrewshire, Glasgow and Lanarkshire. Cottars and other rural labour, cleared from the Highlands and lowland parishes, also found their way into unskilled work, including jobs as navvies and pit labourers. The Mining Commissioner, Thomas Tancred, who was quoted earlier, made disparaging remarks about incoming workers from England and Wales. However, they were not destined for pit labour, but were mainly skilled ironworkers from South Wales and Staffordshire who were imported into the Monklands and the Motherwell area to supply specialist labour for the new ironworks.

Without doubt, incoming Irishmen provided the greatest source of external recruitment into mining in Scotland. The presence of Irish immigrant labour in mining, in the context of strike-breaking, has already been noted for the 1820s. In the following decades, both Catholics and Ulster Protestants came over in a steady stream, and especially after the Great Famine of the mid-1840s, when their arrival coincided with the labour needs of a rapidly developing mining frontier in North Lanarkshire, Ayrshire and West Lothian. 'Ready-made' ironstone miners were recruited in great numbers from the immigrant Irish. In 1841, Tancred had observed that, 'migrating Irishmen commence working underground by engaging as drawers of ironstone, being the business easiest learned'.[17] It was then estimated that a third of the ironstone mining workforce in the Monklands was Irish and by 1850 half of all mineworkers in the area were first- or second-generation Irish.[18] Here, as in other areas of ironstone mining in the west of Scotland, they were employed mainly by contractors for heavy haulage tasks. As adult male labourers working on low wage rates, in the first instance they usually replaced any women and young workers, while some progressed to the supervised longwall squads, where they took down and hewed the coal or ore, or helped to construct and maintain the connecting roadways. William Dixon, Junior, leading coal- and iron-master, and an experienced employer of labour in the greater Glasgow area, compared the character of recruits into mining in the mid-1830s, and stated his preferences:

> The reason why there are so many more Scotch than Irish among the boys, is that they are generally the sons of the colliers, and there are more Scotch than Irish colliers; moreover the Irish are generally younger men, and have not been so long with us.

The Irish in the coal mines with us bear a good character. We have nothing to complain of them on that score; they are fully more obedient and tractable than the natives, and are not so much given to combine; they are lively, and sometimes, when they get drink among them, they are a little excited. They are very much disposed to learn any thing you put them to; they do not find so many difficulties in beginning anything new. An Irishman, who has never seen the mouth of a coal pit in his life, has no hesitation in going down and commencing what you ask him to do. They are, perhaps, quicker at taking any thing new than the Scotch, that is, in the same class.

We find them very useful labourers, and their services are of considerable importance to us; at present, we could not do without them. In this part of the country, the Scotch do not show much disposition for labouring work; they would rather go to trades. A great majority of the West Highland hands are quite useless; in general they are a lazy, idle set; we decidedly prefer the Irish to these Highlanders.[19]

Strict disciplinarians like Dixon welcomed the compliant Irish labourer turned mineworker for both stoop and room and longwall working, and took advantage of the buyer's market for this source of ready labour in the western coalfield. The coal and iron companies used inexperienced labour from Ireland and elsewhere to work the thick seams being opened up in deep pits. From the coal-masters' standpoint, the process of extracting coal from seams which were five or seven feet thick did not require all the skills and judgement of the best practical miners, and it is understandable how established miners continued to resent bitterly the intrusion of untrained interlopers (Irishmen and Scots alike) who threatened to dilute their own hard-won skill and status.

Until the 1870s and beyond, this was the typical employment position in those parts of the west-central coalfield that were dominated by the iron companies. However, a different pattern of employment and recruitment prevailed in the eastern coalfield areas of Clackmannan, Fife, Midlothian and East Lothian. In these counties, one of the crucial differences in the composition of the workforce in mining compared to the west was the greater number and continuity of families of traditional mining stock, as against the minority presence of incomers and strangers. The relative stability of the mining population of Midlothian is especially noticeable. The low mobility of Midlothian colliers is indicated by analysis of census reports, which show up a high incidence of working life in the same districts as those

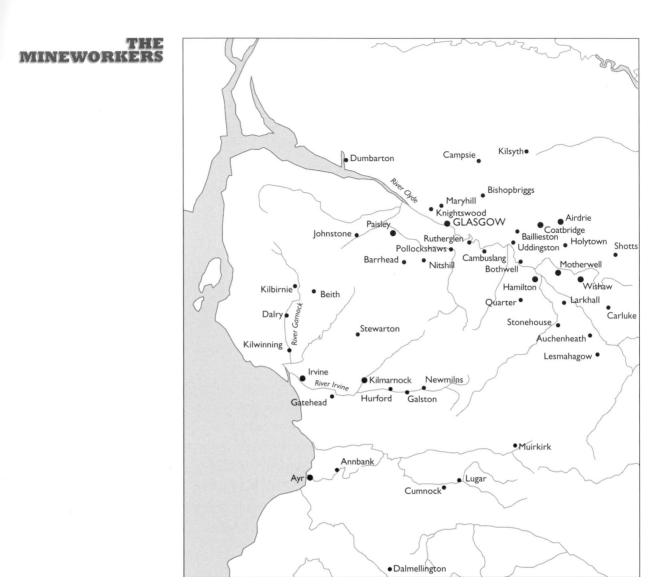

Plate 33.

Principal mining communities in west of Scotland, mid nineteenth century. (*G Wilson, Alexander McDonald*)

of their stated places of birth. For example, in 1851, nearly 70 per cent of all Easthouses colliers had been born in the village and 91 per cent in the county of Midlothian. Similarly, in 1861, as many as 95 per cent of all Newtongrange colliers had been born within the coalfield area. This pattern of stability was partially a reflection of the slower rate of expansion of the Lothian mining labour force and of the industry itself between 1840 and 1880, compared to the faster tempo of growth and extent of outside recruitment within the western mining frontier areas. While it is certainly the case that, after the 1840s, the labour force in the expanding West Lothian mining frontier was swelled by Irish immigrants, incoming Irish penetration of the collieries of Midlothian and East Lothian was almost non-existent. Thus, the colliers of Midlothian remained fundamentally Scottish to a startling extent, and this characteristic is all the more remarkable as Edinburgh was proportionately the second most Irish city in Scotland in 1851. Irish railway navvies were abundant there and in the nearby coalfield in the 1840s, but they did not subsequently move into pit work. Close to Edinburgh, the mining parishes of Newton and Newbattle were virtually free of Irish presence in the 1860s. In 1871, the mining village of Easthouses contained only 1.8 per cent Irish-born and Newtongrange 1.3 per cent.[20] Such high concentrations of indigenous mining stock contrasted with the mixed character of most mining communities in the west, where the Irish presence was a very visible and constant feature of that mix. And while we shall see that the eastern coalfield areas were not spared from industrial conflict in the mid-Victorian period, they were at least relatively free of the ethnic rivalries and sectarian rancour which split so many of the mining communities of west and central Scotland.

Conditions and Grievances

The drive for greater efficiency and higher rates of productivity per underground worker continued to be at the expense of safety considerations, despite some important technical developments in winding, haulage and ventilation, and greater regulation and inspection from the 1850s. The official death toll from mining accidents in Scotland during the 1860s and 1870s makes grim reading, averaging almost 100 a year. Only fishing, stone and slate quarrying exceeded the accident mortality rate in the mining industry overall. The 1871 statistic of one accidental death for every 461 colliers in the country represents a very high fatality rate from accidents at work.[21] Respiratory diseases arising from breathing coal and stone dust, fumes from candle or oil lamps and pit gases claimed an unknown toll of death and debility. Among pit workers in their thirties and early forties, coughing up the 'black spit' and premature death from lung and chest disease were regarded as inevitable facts of occupational life. Taking all these wasting diseases into account together with the myriad of serious non-fatal accidents, the recorded death toll from accidents is only one indicator of the levels of danger encountered and the real cost in life, limb and human misery.

The typical case history of avoidable injury and mutilation can be illustrated from one mining locality in Lanarkshire. Wishaw's thick seams were worked by stoop and room and the Mines Inspector had reported a particularly high accident rate here in the early 1870s. There were too many accidents from roof falls, affecting men and boys alike. This was often the result of inadequate propping and of hurried blasting operations, and

happened mostly when coal was being removed from the stoops or pillars in the final return stage of active operations in a section. This particular task was supposed to involve only very experienced and careful colliers, but this precaution was not always observed. For example, in July 1870, Hugh Walker, aged twelve, a hewer's mate at Greenhead No. 2 pit 'had been cutting down a stoop, and had fired a shot, but the detached mass did not all come down. He then commenced removing some stone from the hanging mass, when it suddenly came away and fell upon the lad. One of his limbs had to be amputated.' On May Day, 1868, John Yuille, described as being 'about 11 years of age', met his death in Muirhouse Pit. The boy, a pony driver, was taking empty hutches to the coalface when part of the roof above the underground road caved in on him. He died within an hour of being carried home. At Sneddon's colliery, in 1865, two men were killed while being lowered down a shaft. The insecure hemp rope that wound the cage snapped, causing an eighty-foot fall. Rail and wagon accidents at sidings and pitheads, and from careering hutches in underground roads, were also frequently reported. Again, many such accidents involved boy workers. The act of snibbling – inserting levers to stop moving wagons – was a dangerous task, and one which was clearly unsuitable for youngsters. Snibbling required careful timing, alertness and strength, particularly when dealing with heavy wagons moving at speed down an incline. In 1870, a boy of ten, unloading coal at Netherton Pit, failed to snibble the wagon, which ran over his legs. Amputation could not save his life.[22] It was the custom to carry home injured and maimed workers, where they were treated by a works doctor, but when further treatment and surgery were required, it was necessary to undertake the agonising and seemingly endless journey to Glasgow Royal Infirmary.

Between the 1840s and 1870s, the most frequent causes of death in a pit continued to be from roof falls, accounting for almost half of lives lost. Despite the increasing replacement of hemp ropes with wire, and the introduction of self-acting safety guides in the winding gear, accidents in the shaft when cages or workers fell down all or part of the depth of the shaft came next in the order of fatalities. Explosions, usually of firedamp, caused the third largest category of accidental deaths, and as some explosions were the cause of multiple deaths, they captured sensational press headline news as 'pit disasters'.

In this period, there were several notable disasters of this sort, all arising from defective ventilation and management failure to enforce the necessary safety precautions, which might have prevented their occurrence. Commonhead Pit, Airdrie, was old, badly designed, with irregular stoop and room workings. It was known to be gassy, and miners had been instructed to waft gas out of the working areas, in the usual crude way, by using their jackets. In July 1849, over a two-day holiday period, ventilation furnaces had been shut down, allowing the methane to build up. Eighteen men were killed after a naked flame from a miner's lamp had ignited the explosion. This was the worst mining disaster to date in Scotland, but it was followed in 1851 by one of even greater magnitude, this time at a modern showpiece colliery and the deepest working mine at the time. Sixty-one men and boys died in the explosion that ripped through the workings of the Victoria Colliery at Nitshill, in Renfrewshire. This episode showed that the skill and under-standing involved in the technical processes of deep mining were not matched by awareness of the need for safe working practices, especially to contend with the movement of inflammable gas at such depths.

However, the greatest pit disaster in the whole history of mining in Scotland, killing 207 men and boys, occurred at Dixon's Blantyre Colliery on the morning of 22 October 1877. The deep stoop and room workings were known to be gassy in this newly developed colliery and the usual early morning safety inspections of the three pits were undertaken by the firemen, who had found nothing abnormal. The official enquiry revealed the apparent causes of the flaming explosion, which, within a few minutes, had roared around the miles of underground roadways. A large ventilation furnace had been allowed to burn low during a recent shift change, affecting the airflow, and a spark or naked light from a miner's lamp had ignited firedamp, which, in turn, had ignited the coal dust in a series of fierce dynamite-like explosions. A survivor, Andrew McAnulty, aged seventeen, who gave evidence to the enquiry, related his experience of an earlier fatal incident, in August, which ought to have provided sufficient warning of the need for proper precautions:

> I was down no.2 pit at six o'clock one morning … with my brother
> Joseph, who had a contract for working out the stoops. I was
> engaged in laying rails with my brother, and while we were so
> employed the gas exploded at my lamp, which was a naked light.
> That was about eight o'clock. Both my brother and myself were
> severely burned. My brother died at ten o'clock that same night in
> consequence of the injuries. I was burned on the hand, arms and
> back. I was working about half a stoop off the level road, eight yards
> from the stoops. All the workmen I saw about the place were using
> naked lights. No one told me to be careful with my light, as there
> was gas about the mines.[23]

Following the enquiry, there was no prosecution against owners or managers for breaches of mining legislation governing inspections and safety lamps. In this instance, as in so many other cases of this kind, the law proved a dead letter. It took this massive loss of life at High Blantyre to bring about an essential safety improvement, which should have been installed in the first place, namely a ventilating fan of 200,000 cubic feet per minute capacity to blow an effective air supply through the deep workings.

In the new shale mines, working conditions and safety regulations were equally poor. In the early years, it was possible to extract the mineral by opencast working or by shallow drift mining, taking advantage of the many outcrops. As the industry expanded, larger quantities of shale were required and a more productive method of extraction was introduced. Numerous faults and folds made it difficult to gain access to large continuous areas of shale from a single shaft or mine. Therefore, until 1877, the sharply inclined seams of shale up to three or four feet thick were extracted by the longwall method, with three-man teams working in cramped conditions. From the end of the 1870s and into the twentieth century, stoop and room was the main method in shale extraction, as it was more suitable for exploiting the much thicker seams, up to ten feet and more, which were then becoming available.

Gunpowder blasting was used more frequently than in coal extraction to bring down the hard, heavy, slate-like slabs, and the constant dust raised the levels of danger from silicosis. The combination of dust and powder smoke also decreased visibility, and made it difficult to keep the air clear. Teams worked with naked lights, and most pits had a furnace installed to assist ventilation. However, shale pits had only one shaft until after the shale mine disaster at Starlaw, near Bathgate, in 1870, in which seven men were killed and others suffered from terrible burns. The single shaft was partitioned into two sections, which served as the only entry and exit, and as a flue for a ventilation furnace at the pit bottom. Sparks from the furnace underneath set fire to the already sooty wooden lining of the shaft. The cage managed to make several frantic journeys, packed with men and boys, while the shaft fire blazed around them, and most were saved before the winding ropes finally snapped. This tragedy drew attention to the absence of official regulation of safety in shale mines and particularly to the dangers arising from having only one means of escape. After the Hartley colliery disaster of 1862, all coal pits had to install a second shaft, but shale mining was not covered by this legislation and full inspection until 1872. As in coal mining, it was only gradually over several years that the more expensive mechanically driven fans replaced the less efficient ventilation furnaces.[24]

Plate 34.

Shale miners at stoop and room working: hewer with pick; drawer shovelling shale into hutch.

(*West Lothian Council*)

It can always be argued that many accidents from roof falls, naked lights, gas explosions, shot firing and other hazards might well have been prevented by miners exercising greater care. However, in mitigation, miners were far from being in control of their working practices and working environment, and safety provisions were not a priority for mine-owners and operators. Further, some basic points of information on the various pressures bearing upon the miner at the working face are required to explain his pattern of behaviour, including some occasional risk-taking. As miners were paid on tonnage rates in coal, ironstone and shale, any time spent, for example, in waiting for changes in the unpredictable ventilation system to clear a working place was unproductive. He was not paid for time lost, for whatever reason, and there was considerable incentive to adopt less orthodox procedures to the extent of taking calculated risks. Working in pitch darkness, the miner chose to see around him by the naked light of a tiny flame from a crude oil lamp, unless he took a safety lamp instead, especially in known firedamp conditions. However, safety lamps had to be carried by hand, had to be laid down and were considered a nuisance to work with. They were not popular, as they did not give out nearly as much light as a naked flame, due to the close mesh gauze that formed the barrier between the flame and the atmosphere. Moreover, as he worked, the hewer had to be sensitive to adverse changes in the working environment. If he suspected the presence of dangerous gas, he would have to leave his place to bring the fireman, who could make a proper examination. This meant the loss of

Plate 35.

Hewer undercutting with pick. (*Collection of the Scottish Mining Museum Trust*)

valuable cutting time, so that the sureness of his judgement in such cases had to be carefully weighed. Occasionally, according to accepted practice, he might take matters into his own hands and he or a workmate would take steps to wave away the pockets of gas from the immediate area.

The same consideration applied to the vital and highly skilled processes of cutting and propping. Both in stoop and room and in longwall, one of the hewer's first tasks was that of 'holing' or 'bottoming', where a deep cavity was cut at the foot of the face, rising two or three feet from the floor. As the primary aim was to remove the coal in the largest possible pieces, he had to cut the deepest possible hole without shattering the upper part of the face, which now hung above him as he wielded his pick, wedged and cramped into the cavity. How far he holed under the coal mass was both a vital safety and earnings issue, as he was paid a lot more for large coal. To avoid serious injury from the risk of the top coal falling down on him, he had to judge how many 'sprags' or small timber props to install to retain the hanging coal in position. At the same time, he had to calculate how often and how long to stop for this safety operation, which meant time and money. Bringing down the top coal in big lumps involved a shot or blast of powder if the type of coal was hard. There were many hazards in preparing a shot, until patent fuses became common in the 1870s. Checking an unexploded shot was a life-or-death risk, and hewers were often killed by failing to wait the recommended time before going to examine a fuse which they thought had gone out, when it had only been delayed.

Teams of mineworkers who were engaged directly by contractors, mainly on longwall operations, were pressurised to work in conditions where safety was often compromised by the drive for maximum output. Known in England as the 'butty system', contracting or sub-contracting in Scotland was used by mine-owners as a means of managing and regulating work discipline and production. In areas of rapid expansion of mining in the middle of the nineteenth century, there was a shortage of adequately experienced and educated managers and oversmen. Paying a middleman to run the mine and control what seemed a raw and unruly workforce was the chosen option for large coal-owning iron companies such as William Baird. The contractor was commissioned to recruit and pay labour for all pit operations, supervise the winding gear and other equipment, enforce colliery regulations, and raise the iron or the coal at a price per ton. In turn, the colliers and miners were paid on the normal piece-work basis of so much output per cart. There was an obvious financial incentive for the contractors, many of whom were former pitmen or surfacemen, to drive a hard bargain with their workers and to get the most out of them by fair means or foul. Contractors were highly resented by the hired hands, as they were notorious for their arbitrary and sometimes brutal behaviour, stamping on any attempts to restrict output and outlawing trade union organisation. Moreover, in many instances, mining contractors in this period also had an interest in truck stores, either owning them or leasing them from the company, and involving their relations in running them. Liquor was nearly always available in the truck stores, and many contractors deliberately arranged payment of wages there, or at a public house. In typical fashion, using threats and intimidation, they were in the habit of insisting that their workers buy goods at the store and obtain credit lines in between pay days. In the worst areas where truck was prevalent, until the 1850s, enforcement of 'long pays' of up to a month and even five weeks ensured that many mineworkers and their families were rarely free from debt and the harassment and humiliation which accompanied this oppression.

In the west, many mineworkers were also being employed as daily wage hands without any contracts, and were liable to immediate dismissal for any reason. In such conditions, there was no place for the independent collier who prided himself on having some control over his pace of work and on his ability to bargain for a reasonable return for his labour. This category of collier could survive only in those workplaces where a rigid, imposed work discipline and systematic abuse was not the order of the day. The large integrated coal and iron companies geared their production in a completely different manner from small masters who were in the sale market, typically supplying local domestic consumption. The sale coal companies with only one or two collieries had a busy winter season and a slack summer one, which made possible a more relaxed working pattern and relationship with the employer. Here, the collier was not subject to the harsh managerial regime of the iron company, which, in a highly competitive market depending on cheap labour, required a huge and continuous supply of coal and ironstone for as long as the furnaces were in blast throughout the year.

However, in all collieries where expensive winding gear had been installed and regular output was required, it was difficult for miners to control the number of hours they spent underground. The collier, who formerly had scope to enter or leave the pit at his own pleasure, was now obliged to wait until the appointed time to ride in the cage. He might have hewn all the coal necessary to make his wage for the day, but until the signal had been given, he had to wait at the pit bottom, tired, sore and damp with sweat. Much depended on the policy and practice of the manager, contractor and oversman, and their relationship with the worker, as they controlled winding times and movement of the cage, and could impose long hours underground. Until the 1870s, there are reports of miners being below ground for up to twelve and even fourteen hours a day, and even strong, fit miners could not work physically for more than eight hours in a day. In deep and extensive pits, much of this time was spent walking or crawling from the pit bottom to the coalface, a journey of up to a mile, and back again at the end of the working day. In a large mine with a lot of shift workers, the time spent lowering and raising the workforce added to the length of the working day. It is hardly surprising that working hours and related earnings remained contentious issues. The operation of company stores and truck has already been described for the early nineteenth century. Between the 1840s and 1870s, this system of daylight robbery and abuse of the workforce was practised extensively in all the mining counties of west and central Scotland (including West Lothian), and particularly in those localities where the large iron companies held sway. According to a survey prepared by the miners' leader, Alexander McDonald, in the 1860s, 16,000 workers – half of the workforce in mining in Scotland – were subjected to the truck system at their workplace; and most of those miners were employed by iron companies.[25] It was a major grievance among colliers and miners, but difficult to resist and oppose, when companies and contractors so often had the upper hand in hiring and firing.

Analyses of the conflicts and long-standing grievances endemic in mining during this period reveal that the ostensible cause of most disputes concerned wages and earnings. The coal and iron sectors within industrial capitalism were subject to violent booms and slumps every few years with fluctuations in market demand and in prices. In slump years, companies sought to keep down production costs while continuing to compete with each other in a falling market. Faced with declining profit margins, they cut costs by screwing down wages, imposing short-time working and redundancies. Colliers and miners employed by iron companies had to face massive wage cuts in years of severe crisis, such as 1842 and 1874, provoking major industrial conflicts, which will be discussed later.

One of the fundamental sources of discontent revolved around the weighing methods of the mine-owners, and it is reckoned that, second only to direct wage cuts, disputes over unfair weighing provoked most other strikes in the west of Scotland between the 1850s and 1870s. As hewers were paid by piece-rate, accurate weighing of their output was vital if wages were seen to be consistent and fair. Only one method stood up to scrutiny, and was used in only a few mines. In this simple instance, each hutch was carefully weighed at the pit-head and payment made according to the weight produced. However, several methods of weighing hutch loads were fraudulent. Unjust weighing was extensive, and this could take several forms.

A miner was unfairly penalised if he sent up hutches with different weight loads, and was paid only at a rate for the lightest one of the batch. Elsewhere, a miner lost payment if he sent up a hutch below the declared standard weight, whereas no extra payments were made for hutches over the required weight. One malpractice involved contractors insisting that hutches filled by any of their team had to be heaped up to ensure that the weight reached the required limit. The difference was pocketed by the contractors in collusion with the pit-headman. Another method involved an arbitrary judgement being made by the pit-headman weigher as to whether or not hutches were up to standard in both weight and mineral content. The miner or collier was often unfairly penalised for filling hutches containing too much stone and blaes attached to the coal or ore. A hutch allegedly failing to meet this weight and content was declared a condemned load, and no payment was awarded. Again, contractors and weighers were accused of conspiring to split the proceeds from the confiscated hutches. An analysis of the evidence from the mineral accounts of one famous owner reveals how unjust weighing affected miners' wages in the 1840s and 1850s. In the Duke of Hamilton's six Lanarkshire collieries, statements of production and sales showed that more coal was sold than the miners were paid for producing, even allowing for the additional stock remaining at the start and end of each year and accounting for those hutch loads that were deemed to have excess stone. Whatever dubious weighing practices were being pursued in these collieries, the overall amount of unpaid tonnage was such that miners were being swindled out of one in every eighteen tons, or a deduction of around 5.5 per cent from their wages. In the early 1840s, the 'free' coal yielded for the Duke almost £1,000 per year, at a time when most colliers were earning rock-bottom gross wages of less than 3 shillings a day and a much smaller take-home pay.[26]

Although the Mines Act of 1860 granted miners the right to appoint their own checkweighmen, or 'justicemen', in each pit to ensure fair and proper weighing, many employers resisted this change. And while miners still complained about faulty weighing machines, even as late as 1870 it was officially revealed that mine-owners in Scotland were opposed to installation of a uniformly accurate weighing system because of the alleged expense of the machines and the wage-bills of the worker-appointed checkweighmen.

In Lanarkshire, on the eve of the 1842 general strike among miners, another unfair practice that angered colliers had recently been imposed. Colliers used a riddle for separating dross from coal lumps before filling the hutch, and were not then paid for dross anyway. They found themselves further defrauded when the mesh spaces were widened to 2.5 inches.

Apart from the grief caused by wage cuts, truck, long pays, unjust weighing and indebtedness, net wages were low in most years, as the earnings of mineworkers were also subject to compulsory reductions, or 'off-takes'. These non-negotiable reductions could result in a loss of as much as 20 per cent of gross wages and discriminated against many workers. Colliers did not object to the main off-takes such as house rent, house coal and tool maintenance, as this was an accepted arrangement. What caused resentment and discontent was the aggregate loss of earnings involved, and the number of deductions in some workplaces. For example, at Dixon's Govan collieries there was no truck and pays were fortnightly, but a tight managerial regime was preferred. Several separate deductions were imposed, including money for house coal, pick sharpening, colliery school fees, subscriptions to the company's friendly society, funeral fund and services of a company doctor.

Moreover, here and elsewhere, the amounts paid often depended on the individual circumstances of the mineworker and the policy of the employer. Off-takes payments disadvantaged the steady hard-working hewer because he tended to need more pick sharpening, lamp oil and powder. The larger a man's family, the more school and doctor's fees were liable, and occupying a bigger company house meant a higher rent. Thus off-takes could contribute to the impoverishment of mineworkers and their families through no fault of their own. In such circumstances, desperate to make ends meet, it is little wonder that the wives of mineworkers in places like Coatbridge, where workers' housing provision was scarce, took in so many single men as lodgers, and 'double-shifted' their beds as one worker alternated with another for work and rest periods.

Single pitmen had to pay the levy for the colliery school, although they had no children to attend it. There was also a special grievance experienced by Irish Catholic workers who were compelled to pay for Protestant teachers and religious instruction and for doctors whose medical advice and practices offended their religious persuasion and beliefs. The company appointed the schoolteachers and doctors, and there was no right of objection or exemption for any of the workforce, who had all to contribute towards the costs of their salaries and services.

All these impositions appear to indicate that many large employers, whether iron companies in the western counties, or traditional landed coal-masters as in the Lothians and Fife, were alike in seeking to control the mineworkers not only at the workplace but also in their private lives and at community level. During the mid nineteenth century, large mining employers such as the Bairds, the Fifth Duke of Buccleuch and the Eighth and Ninth Marquises of

Lothian, built or subsidised increasing numbers of schools, churches and houses, and encouraged respectable cultural and social activities aimed at civilising their workforce in the company-controlled towns and villages. The various ideas and schemes, which were designed to improve work discipline, and instil respectable patterns of behaviour and social improvement, may usefully be classified under the general heading of 'Victorian values'. These values encompassed cleanliness and godliness, the virtue of hard work and its rewards, thrift and wise spending of wages, self-reliance, sobriety and deference towards social superiors. Obviously, any defence of worker interests, by means of trade union or industrial action, was anathema to this employer-sponsored scheme of values.

Of course, most mine-owners did not bother to promote such concerns, particularly those owners who did not have the resources to pay for schemes of improvement, and who would not consider investing in housing of a tolerable standard for their workforce. Even some of the large iron companies, for example, Merry and Cunningham, in Lanarkshire and in Ayrshire, did not display a positive attitude towards influencing their workforce, apparently being somewhat negligent of their welfare. For decades, they operated the truck system because it was highly profitable, even though its abuses deterred good workers from staying with the company and was a constant source of unrest. This type of employer preferred to use crude coercion to discipline the labour force.

From the 1840s, the paternalist stance of the very wealthy Duke of Buccleuch encouraged sobriety and deference among his workforce in his lead-mining settlement at Wanlockhead, permitting them to build and improve their houses and extend their smallholdings. In a low-wage regime in this remote hillside setting, as also at Leadhills, owning livestock and vegetable produce helped to protect mining families from destitution in periods of hardship. At Tyndrum, between 1838 and 1862, the supposedly paternalist Marquis of Breadalbane pursued a scarcely enlightened policy towards his estate colony of lead miners. The enterprise was run much like a public works scheme to provide employment, on subsistence wages, for distressed and displaced crofters, although the mine managers were German. The workers lived in miserable hovels, on a diet mainly of vegetables and whisky. A company store sold cheap liquor, and this demoralised and deprived mining population was without a doctor and school.[27]

In contrast, those pit-owners who made a concerted effort to mould the behaviour of their workforce perceived investment in colliery housing as a means of providing accommodation and improved living conditions. It was also an instrument of control and intimidation. A rented house tied to work meant that owners could use the weapon of eviction against workers who defied them in any way, such as breaking colliery rules, taking strike action, or disobeying company regulations governing their behaviour as tenants. For example, a ban on keeping animals such as dogs, pigs or poultry in the house or on the adjacent premises was often imposed, and considered a necessary step to encourage a clean domestic environment. The eviction weapon was ruthlessly exploited by all owners, especially during strikes, and even the threat of eviction was often enough to induce compliance with the employers' demands.

However, although many owners provided houses for their workers, very little of this stock was of decent or model standard for the time. Typical colliery housing of the period was inferior and miserable, often brick-built, low and single-storey, without a damp course. The traditional plan of mining community houses in Scotland was essentially the single room, and nearly always single-storey. They compared poorly in space and comfort to the general stock of colliery housing in England and Wales. With slate or tarpaulin and wooden roofs, the houses were set out in rows or squares, with outdoor communal taps for water supply and open drains and ash-pit middens to dump human and other waste. Small, cramped, overcrowded, ill-ventilated and without coal sheds or shared wash houses, they encouraged neither cleanliness nor comfort.

Plate 36.
Low, single-storey miners' rows, New Stevenston, Lanarkshire. (*North Lanarkshire Council*)

A better standard of housing was built in the two-storey tenement style in some of the mining towns and villages. For example, an angry Wishaw miner in 1857 exclaimed that 'the coal masters' horses have more comfortable habitations', referring to some of the worst houses in the area – wet with damp, one-roomed, with no back doors, closets, or privacy for whole families. However, he pointed to new houses going up in the nearby company village at Newmains, where Mr Houldsworth, iron- and coal-master on the Coltness Estate, had ordered roomy, two-apartment buildings with living room, bedroom and scullery. Houldsworth had already built such improved houses for his pit workers at Dalmellington.[28] These were among the few exceptions to the provision of generally poor-quality housing and amenities in mining areas. The most notable exceptions were in Midlothian, above all the new colliery houses at Dalkeith, built in 1840 by the paternalistic landlord, the Fifth Duke of Buccleuch. Here, although a low-wage community, at least there were no truck stores and no contractors, and the colliery houses were rent-free. They had large rooms, were well lit, had an excellent water supply and, almost uniquely for colliery houses, were furnished with individual water closets.[29] This high quality of colliery housing was rarely matched elsewhere in Scottish mining districts, but where better housing stock existed in any mining community it was intended to attract and reward the most reliable and compliant workers, notably the steady and respectable married men with young families.

In their civilising mission, masters like Henry Houldsworth combined the provision of decent housing with a strict policy of prohibiting licensed premises on their estate or company village. At Newmains and other pit villages on the Coltness Estate around Wishaw, and at Dalmellington, Houldsworth set an example and no shop or company store was allowed to sell hard liquor. Such interventionist employers were determined to stamp out drunkenness and disorder among their workforce, not least as such bad habits interfered with work discipline and production. From the 1840s, often in conjunction with local churches, they promoted temperance and teetotal campaigns among the workforce to bolster the concern for sobriety.

This assault on the behaviour of mineworkers did not stop there, as the gospel of improvement extended to other activities. The policy of the Bairds at their Coatbridge base involved a whole range of provision, including schools, houses and churches, reading rooms and sports halls, baths, workers' institutes, lectures and concerts, a total abstinence society and a colliery band.[30] The wealthy aristocratic mine-owners at Newbattle and Dalkeith also ploughed money and personal effort into various schemes of moral and social improvement, all intended to promote social harmony, good

living and loyalty among the workers. Their missionary and temperance
work was allied to high-quality housing and a generous supply of colliery
schools, reading rooms and subscription libraries. However, in common with
all other mine-owners, despite displays of paternalist concern and giving,
they would tolerate no challenge to their authority and power, and workers
did not have to be reminded who was the master in the workplace and in the
mining community.

Mineworkers: Responses and Resistance

Throughout this period, as in earlier times, the dominant image of
mineworkers, as portrayed by outside observers and commentators, is
overwhelmingly negative and hostile. Whether considered as individuals or en
masse, miners are variously characterised as ignorant and illiterate, brutish
and uncivilised; and comprising an underclass of black savages who were to
be either pitied or feared. However, there is also ample evidence to counter
such a prejudiced image, and to demonstrate that it is a misleading and one-
sided stereotype.

   To divide miners into 'rough' and 'respectable' is a useful, though
incomplete, distinction. The drunken, improvident, betting and gambling,
feckless miner, who also preferred the traditional and brutal blood sports of
cock-fighting and bare-knuckle fighting, and spent time and money in illicit
drinking dens and other unsavoury pursuits, was obviously evident. And
whatever the deficiencies of his personal character, the adverse circumstances
of this type of miner were usually worsened and his outlook further
demoralised by the weight of the abusive conditions outlined above. In most
mining communities, there was an obvious choice between the contrasting
values and culture of church and chapel on one hand, and those of the public
house on the other. However, irrespective of the severity of their working
conditions, it would appear that many more miners were neither consistently
respectable nor rough in their everyday behaviour and lifestyles. If temporary
solace and escape were all too often found in resort to cheap liquor, the
leisure pursuits of miners at this time and later also reflected a desire to free
themselves from the bonds of underground labour and to see much more of
the sky. Hence, the popularity of outdoor activities such as keeping dogs and
whippet racing; pigeon fancying; playing quoits; gardening, if available;
amateur football and cricket, and the occasional poaching expedition for
fish, rabbits and other game, at the risk of falling foul of the law.
Nevertheless, it was the sober, the self-reliant, the thrifty and the serious-
minded miners who can be perceived as occupying the positive, activist end

of the spectrum. Moreover, they were generally the ones who believed and persevered in the notion that individual willpower and collective organisation could alleviate hardships and win redress of grievances. Such values also embodied the fighting spirit of the independent collier, determined to defend his craft and a living wage against the constant threat from unskilled labour, and the deterioration of standards and conditions. This stance of the respectable men was reflected in their leadership and support of friendly societies and trade unions. Their self-improving aspirations also led many to participate in further education and rational recreation, to subscribe to adult classes and lectures, reading rooms and libraries, and to promote temperance and teetotalism to counter the baleful effects of the demon drink.

Alexander McDonald and, after him, the young Keir Hardie, were exceptional characters within a minority of miners who upheld such values. Both men often railed against drunken behaviour among miners, and in the late 1870s, Hardie did so as a prominent member of the Independent Order of Good Templars, who campaigned for the statutory closure of public houses.[31] Moreover, in taking up their principled stance, the representatives of the miners were vitally different from those employers and middle-class elements who also supported such values. They did so as trade unionists, passionate about the justice of raising the living standards of their class, pursuing the objectives of working-class improvement for the good of the miners and their families.

Even before the expansion of self-improving societies among better-paid sections of workers from the 1850s onwards, there is ample evidence that significant numbers of the more prudent and provident miners formed and sustained friendly societies to protect themselves and their families from hardship and the humiliation of poverty. Over twenty friendly societies in West of Scotland mining communities alone were officially registered between 1829 and 1852. There were many other similar societies that did not register, including those that functioned on an annual basis, dividing up the funds among their members at the end of the year. The yearly societies were more popular among miners who were frequently on the move from place to place. The annual and more permanent societies alike served as savings banks and as a means of social security against sickness and injury. Exercising thrift and self-help in this way was also designed to provide a measure and feeling of independence. For instance, the Airdrie Miners' Friendly Society of 1832 existed to fund illness, injury and funeral benefits for family members, and the running of such societies was entrusted to sober, reliable and socially committed miners and sympathisers. The Woodhall Colliery Friendly Society, near Coatbridge, insisted that the treasurer could not be a spirit dealer. Neighbouring friendly societies in Calderbank, Chapelhall and Whifflet were glad of the services of local schoolteacher, James Hair, as secretary. And the able, articulate Irish-born miner, William Cloughan, leading figure among the Monklands miners in the early 1840s, had been secretary of the Thankerton Colliery Friendly Society for several years previously. As models of collective organisation with rules, structure and funds, the friendly societies run by committees of miners and other working men provided a training ground for potential leaders and officers of trade unions. They also encouraged miners to attempt to provide similar welfare benefits as part of the functions of their emerging branch and district unions.[32]

Respectable or rough, miners of all shades so often found themselves in open conflict with their employers over basic grievances, in attempts to secure a living and gain some elementary justice. Even the normally compliant and relatively cultured Leadhills miners confronted their employers in a wages dispute in 1836. The whole workforce of the Scots Mines Company – 200 men – had formed a union and all were on strike. Troops were called out from Hamilton to keep order, although there was no rioting or unruly behaviour, and many miners were dismissed for daring to take strike action.[33]

Between the 1830s and the 1870s, all the major confrontations in the Scottish coal, lead and shale mines were provoked by savage wage cuts. In the many small-scale disputes, at pit or district level, the ostensible cause of strikes among coal and ironstone miners was always pay related, and some were also specific and concerted protests against unfair weighing.[34] In the large coal-sector conflicts, as in 1842, 1856 and 1874, remarkable levels of mobilisation and sustained strike action were achieved among the miners, such was the will and the capacity to resist the power of the employers and, at times, the presence of police and military force. As the 1842 episode exhibits all the features of social unrest and class war in an industrial frontier society, it is given the fullest coverage.

Following the defeats of the 1820s, formal organisation barely existed among the miners in the western counties, and isolated acts of brutal intimidation of strike-breakers, including ear-cropping, marked the desperate weakness of resistance to employer control of labour. In June–September 1837, trade depression and wage cuts provoked extensive strikes and employer lockouts in Lanarkshire and the Falkirk area. Sheriff Alison called out troops to protect strike-breakers and quell disorder, in his case setting a pattern for law enforcement in serious mining disputes throughout his long career as principal lawman.

In early August 1842, 12,000 miners in the Monklands and Glasgow districts took mass strike action, coordinated by an informal network of local strike committees. By mid-August, the strike had spread to parts of Ayrshire, and to Midlothian and East Lothian. This extensive action was significant in Scotland and beyond, taking place at a time when textile workers and mineworkers in other industrial areas of Britain were also involved in large-scale strikes and political protest. The August strikes included nearly 500,000 workers, comprising the largest mass mobilisation of its kind in nineteenth-century Britain.

Although the striking workers, including the Scottish miners, were primarily concerned with wages and conditions issues, in their thousands they were also raising political demands that challenged the whole system of power in the country. Interested in securing democratic rights, they supported the movement for the People's Charter, and its six points, which included the vote for adult workers and representation of working-class interests in Parliament. In the Monklands alone, Chartist branches existed in Airdrie, Coatbridge and Holytown. McLay, secretary of the Airdrie miners, was an ardent Chartist who supported the call for an all-out political general strike until the People's Charter had become the law of the land. Although this extreme demand was a minority position among miners, it was also voiced by meetings of Fife and Clackmannan workers who had refused to take strike action on workplace issues alone. It was the fusion of workplace grievances and political demands, supported by mass protest throughout the main industrial districts, that created a potentially explosive situation and prompted the Tory Government to declare a state of emergency.

In the principal trouble spot of the Monklands, Sheriff Alison had no doubts about the purpose of the miners' strike action. He was convinced that Chartist political agitators were the prime movers and that, in any case, as in 1837, a prolonged strike would result in the breakdown of law and order.

During the first three days, the miners were fully in control of the North Lanarkshire industrial frontier, picketing and patrolling the roads, holding mass meetings and parades. They had gambled on a short, all-out strike to win concessions on hours and payment and were without an organised strike fund, which weakened their position and contributed to the ultimate failure of the strike. Moreover, by breach of work contract, their families were outlawed from poor relief, and the destitute mining community of around 70,000 people had to resort to stealing, begging its food supply and making credit deals with willing and reluctant local shopkeepers. Under cover of darkness, gangs of men, women and children looted the potato and turnip crops from farms. Vegetables and fruit were stolen from gardens; livestock, including sheep and pigs, were lifted; bread and milk carts were seized. All the company stores were closed and boarded up, but the large Dundyvan truck store was ransacked in a night-time break-in.

From the start, liaising closely with the Home Office and the Lord Advocate in Edinburgh, Alison applied the state of emergency in Lanarkshire, offering to mediate between masters and miners, and mustering the means of law enforcement. In scenes akin to those of the Wild West, the sheriff first attempted to raise a posse of volunteer special constables from among farmers, tenants, householders and shopkeepers. However, the response to his appeal for a posse of yeomanry was poor, as only a few estate tenants could be persuaded to join up, and Alison had to act quickly to mobilise professional police and foot-soldiers from Glasgow. By 4 August, additional army reinforcements from Edinburgh, a troop of cavalry and three companies of infantry, had arrived in Airdrie. Alison led night patrols of cavalry to hunt down suspects in the mining villages, but often found that their searches were in vain, and few arrests were made for stealing, assault and intimidation. The punishment for potato stealing was sixty days' imprisonment. This was also the sentence for six Dundyvan miners and five Drumpellier miners, tried in early August for breach of contract, after their employers had lodged a case against them as an example to their other workers. These convictions, with the threat of more to follow, did not have the desired effect of cowing the miners into submission, and food raids on farms and fields continued until the end of August. By September, several iron companies had conceded the four shillings a day demand and agreed to fortnightly pay settlements, which split the solidarity of the strike action and led to an increased physical intimidation of colliers and ironstone miners who had returned to work. No one was reported maimed or killed here or anywhere else, but Alison and the Scottish courts were determined to inflict the most severe sentences, short of the death penalty, against violent intimidation. In Ayrshire, three colliers were sentenced to ten years' transportation, and the same fate befell six Airdrie miners who had participated in a successful siege of the local prison headquarters, releasing several miners who were in custody there. The drift back to work was over by October, mainly on the employers' terms.

Elsewhere, troops and special constables were deployed in Clackmannan and Dunfermline; and in Midlothian, where the strikes lasted longest at the large estate collieries, police and military were called out to protect blackleg labour. Despite their paternalist reputation, the landed mine-owners here did not hesitate to enforce wholesale evictions, and to conduct a systematic blacklisting of strike-leaders.[35]

For the Midlothian colliers, the reality of being starved back to work and the defeat of strike action was a bitter lesson, and for years they resolved against strike action as a means of redressing grievances, even when this meant agreeing to abide by the rules and regulations of the colliery owners. When trade unionism revived in the early 1850s, and a district union was formed, its stance was decidedly moderate, preferring conciliation and relative collaboration with employers, who began to respond by extending improved housing and other welfare and recreational provision in the mining villages. Industrial peace largely reigned until the early 1870s, by which time the principal landed coal-masters were deciding to lease their mineral reserves to iron firms and limited companies. In the Lothians, as elsewhere, the rising demand for coal in the economic upswing of the early 1870s triggered off a revival in collier militancy. The mining community shook off its deferential attitude and again joined the fight for a greater share of the proceeds of a highly profitable market. By the end of 1872, the Mid and East Lothian Miners' Association had 2,000 members, although both organisation and living standards were lost in the aftermath of the ill-fated 1874 strike. Outraged by the recent strong involvement of his colliers in militant trade unionism, the Duke of Buccleuch ended his long-term concession of free housing and soon thereafter abandoned direct ownership and control of his local collieries.[36]

The attitudes and circumstances of the Fife and Clackmannan miners were in the same mould as their Midlothian and East Lothian counterparts during the mid-Victorian period. Here, in general, after 1842, employer–worker relations in mining were less troubled than in the western counties where, in contrast, truck, contracting, payment grievances, high labour turnover and severe fluctuations in prices and earnings were distinctive features. The miners in the counties named above stayed out of the extensive strike of early 1856, which involved around 20,000 miners across all the other districts. This major action was a baptism of fire for the newly formed Coal and Iron Miners' Association of Scotland, which had its main membership base in Lanarkshire and Ayrshire. Weaknesses of organisation, funds and leadership were firmly addressed in this critical strike, and remarkable levels of unity and discipline were achieved, given the sheer numbers involved. The largest police and military presence to date had been called up in anticipation of massive disturbance and disorder, and although intimidation of strike-breakers inevitably involved serious incidents, Alison and fellow law officers agreed that the fourteen-week-long action was conducted in a relatively orderly fashion. They soon concluded that the miners were adopting a more peaceful policy, and the 1860 dispute in the Clyde valley area was the last occasion when troops were used against striking miners in Scotland in the 1842–74 period. In the end, strike funds were inadequate to sustain further action in the long dispute of 1856, and, in most cases, miners had to accept the reduced average wage of four shillings a day.

Building a union with sufficient funds, setting out campaign demands to redress a range of grievances, maintaining unity and discipline and providing firm leadership to take on hostile employers and political opposition, were huge tasks for aspiring trade union leaders. Yet, against the most formidable odds, the accomplished Alexander McDonald, who led the Scottish miners from the mid-1850s until the 1870s, had some success in realising those objectives. A Liberal in politics, his ideals embodied the respectable tradition of the independent collier, and he acted on the conviction that the interests of the mining community had to be articulated, defended and represented at all levels, from the pit to the British Parliament. He urged miners to be responsible and law-abiding at all times, while being resolute in their determination to win justice. He took miners' grievances before Parliament, and persuaded many workers of the value of campaigning on a number of pressing issues. They included enforcement and improvement of existing regulations on health and safety such as ventilation, and legislation on the training of mine managers; on shorter working hours; uniform weighing and an end to the truck system.

After 1856, although only among a minority of miners where union and workplace solidarity was sufficiently strong, selective strike action on wages issues and restriction of output (the short darg) was often applied effectively in local and district disputes, backed up by support funds and levies. In 1870, the Fife miners gained an eight-hour day (a famous victory celebrated by annual gala), achieved by a stay-down strike at a time when demand for labour was high. Yet this was one of the few small victories amid a welter of unresolved grievances. McDonald and others had agitated against truck for many years, forming anti-truck committees to expose the abuses and the masters and store-owners who defended and profited from the system. Little was achieved until a major parliamentary enquiry into the workings of the truck system took evidence in 1870. Public exposure of abuses and campaign action by miners, prominently in Lanarkshire, resulted in the closure of many truck stores. Miners and their families were among the most stalwart members and supporters of the emerging co-operative societies soon to be so closely identified with working-class benefits and improvement. Even then, the evils of payment in kind through compulsory use of the company store were not effectively ended until government legislation in 1887. Although by no means the worst example of employer persistence in this regard, in the

1870s and 1880s, the 700 inhabitants of the four miners' rows at the isolated Benwhat village in the hill moor above Dalmellington were still obliged to shop at the company store. Strong iron chains were set across the approach road over a mile away to prevent the entry of horse-drawn grocers' or butchers' vans, and it was many years before this restriction was removed.[37]

The few large-scale strike actions across the coalfields ended in failure, as did attempts to build a Scotland-wide trade union. Several well-organised, district-wide trade unions existed for short periods, especially in the boom of the early 1870s when wages were at a record high level. Most district unions were located in Lanarkshire, namely in Wishaw, Motherwell, Carluke, Larkhall, Hamilton and Stonehouse. However, they were conspicuous by their absence in the Monklands area and in neighbouring Ayrshire. Since the 1850s, the weakness and sometimes non-existence of trade unionism in areas such as Coatbridge and Dalry reflected the dominant grip of the ironmasters, whose repressive anti-union policies and strategies were largely successful in subordinating the workforce. Here, for many years, the long darg and low wages were the lot of this class of workers who were described as 'degraded and willing slaves'.[38]

Moreover, in such areas, disunity also resulted from internal division among the miners, already occasioned by rivalry over jobs and conditions, but exacerbated to a significant extent by sectarian friction between Irish Catholic settlers and Scottish and Irish Protestant and Orange elements. The Free Colliers' movement, which thrived briefly during the low ebb of trade union morale and membership in the middle 1860s, was an attempt to assert and defend the influence of the Protestant Scots collier. Most lodges were established in Lanarkshire, Ayrshire and around Stirling, but also extended to the east and Midlothian, where there were over 1,000 members. It was not a revival of the earlier secret brotherhoods of colliers, although it was part trade union and part masonic, with regalia and ceremonies. However, it was a divisive influence, as it had links with Orangeism, its outlook was anti-Irish and anti-Catholic and it had the effect of alienating many Irish mineworkers who were obviously discouraged from participation. In many communities, and not only in west and central Scotland, workplace solidarity and unity were difficult to achieve in the face of occupational, ethnic, religious and political differences, producing tension which occasionally erupted into communal riot and violence.[39]

The onset of trade depression and defeat of the 1874 strike broke up the fragile solidarity of the district unions, including the Shale Miners' Association. The main workplace and trade union strategy of restricting output to raise wage levels, reinforced by strike action as necessary, was in

Plate 37.

Opposite. Alexander McDonald.

Plate 38.

Overleaf. Group of miners, late nineteenth century; colliery unknown. Note open-flame oil lamps fixed to bonnet.

ruins. On this critical occasion in 1874, McDonald had advised against a widespread strike on the grounds that wage levels would not be restored while the market price of coal and iron had declined. The conflict at local level is illustrated by the experience of mining communities in Motherwell and Wishaw. The iron company collieries of Merry and Cunningham at North Motherwell and Craigneuk were targeted for strike action, as trade union influence was weak at the Coltness and Glasgow Iron Company pits. Strikers were supported by union levy. The large-scale coal-masters and iron-masters retaliated with a concerted lockout of miners in both localities, thus provoking an escalation of the dispute. Some masters used the customary eviction weapon. Merry and Cunningham secured eviction notices for Logan's Rows in Motherwell and Merry Square, Craigneuk. The families at Logan's opted to leave instead of being forcibly evicted, and the managers of the neighbouring Braidhurst colliery warned their own workers against taking them in. At Craigneuk, neighbours sheltered some families in their already overcrowded single-end and two-room houses. For several weeks, evicted families, comprising over 100 people, camped in a large tarpaulin tent behind the local school, with makeshift kitchen and washing facilities. Eventually, they were beaten by driving rain and destitution.

In the course of the sixteen-week dispute, both district unions exhausted their funds in strike pay, and not until 1879 and 1880 were they able to revive and resist further wage cuts and long working hours. Even then, nothing had changed to improve the appalling realities of class conflict in mining. In August 1879, families were evicted from Logan's Rows and the threat of eviction in other company houses throughout the Motherwell area forced a return to work in the 1880 dispute. There were more than the usual allegations of violent picketing, assault and intimidation during this strike and Motherwell and Wishaw miners and their womenfolk were convicted on all these accounts. Two miners were sentenced to fourteen days' hard labour, and the wives of three miners from the Berryhill Rows, Wishaw, were imprisoned for two days for throwing stones and slag at blacklegs and a mine manager. A group of women from the Camp Rows, Motherwell, were also tried at Hamilton, and were discharged with fines and warnings for harassing strike-breakers.[40] Reports of direct involvement of women during such bitter and often violent disputes are rare, but large groups of women from the miners' rows at Dundyvan had intervened to attack strike-breakers in the 1842 confrontation, and colliers' wives were arrested for the same type of offence in 1874.[41] Such incidents are a reminder that it was the women of the mining communities who bore so much of the burden of strikes and evictions, and who were seen here in a supportive role, even though the sources for this period convey little record of their suffering or their response.

Self-reliance and voluntary restriction of hours and output to raise prices and wages were no longer effective means of collective action and had outlived their usefulness as main planks of workplace discipline, solidarity and union strategy. New policies were needed, and from the 1880s a fresh generation of mining activists, including the young Keir Hardie and Robert Smillie, would begin to raise more radical demands and solutions, derived from labour and socialist beliefs. These proposals would still cover strike action to defend basic interests, but would also advocate direct state intervention beyond such specific aspects as mining safety and child labour, to encompass fundamental demands for a legal eight-hour day, a minimum wage, nationalisation of the mines and state-insured pensions from mineral royalties.

[5] Toil and Trouble in the Peak Years of Mining, 1880s–1920

The Mining Sector in Scotland, c.1880–1920

In total output and numbers of workers employed in Scotland, coal and shale mining continued to expand throughout this period, whereas ironstone-mining and lead-mining began a long decline. As in previous phases from 1700, the growth in coal-mining was spectacular. Despite some bad years, production and the overall workforce both doubled from the 1880s. By 1913 the coal-mining industry had reached its all-time highest level of production. Output levels were disrupted during 1914–18, when vital coal exports from Fife and the Lothians to northern Europe were cut off and as many as one in four Scottish miners had enlisted in the armed forces. However, in the short post-war boom period, mineworkers were confirmed as the largest single category of industrial workers in Scotland. Numbering 146,000 workers (including around 2,000 female pit-head workers), they formed over 10 per cent of the occupied working class.[1]

Into the early twentieth century, the Scottish coal industry was still concentrated in the west-central coalfield, which included the great Lanarkshire heartland, Renfrew and Dumbarton, the Stirling–Falkirk area and West Lothian. In output and manpower, the Ayrshire coalfield had recently been overtaken by both Fife and Clackmannan and Mid and East Lothian, although Ayrshire had more working pits and collieries than each of those smaller and more productive coalfields. Lanarkshire, the most highly populated county, and the hub of the west-central coalfield, retained its Scottish-wide lead in the number of collieries and mineworkers. Here, up to 60,000 workers, incorporating a third of the male working population in the county, were employed in mining before the crisis years of the 1920s and 1930s.

Before 1914, the visibly expanding coal-mining sector in the west of Scotland appeared to be in a prosperous and progressive condition, meeting the demands of its many customers. Its hard splint coal still fuelled the iron industry and more so, from the 1880s, the furnaces of a rapidly growing steel industry, particularly in North Lanarkshire. Its steam coal fed the gigantic railway system as well as an expanding export market. In the promising deep-seam district of the middle Clyde valley, especially between Cambuslang and Larkhall, there was heavy investment in new power machinery and equipment for more efficient sinking, winding, pumping, lighting and ventilation, cutting and conveyance of coal. Mechanical coal cutters started slowly in Scottish pits during the 1890s but, by 1913, 900 were in operation, of which around 500 were in the west of Scotland, raising around a quarter of the region's total output.[2]

However, warning signs were already present about the performance and potential of mining in the west and central coalfields. Before 1914, the greatest concern in the industry was the declining prospects and profitability of the older sections of the north and mid Lanarkshire coalfield. Parliamentary enquiries and official reports confirmed pessimistic impressions of the working out of its thick and most productive seams and the near exhaustion of ironstone. Owners complained about the increased labour costs of winning coal from thin seams; and heavy-industry consumers who were used to ready supplies of cheap fuel also began to count the additional costs.

The western coalfields had produced 80 per cent of Scotland's output in 1870 whereas, by 1913, their production level had fallen to just over 50 percent of the total. During those years, despite the overall expansion of

PLATE 39.
Pit shankers, after completing shift at Woolmet Colliery, 1898. (*Collection of the Scottish Mining Museum Trust*)

output and employment in the west, and the existence of several large and medium sized, modern mines, notably in mid Lanarkshire, the best prospects for profitable coal-mining were located elsewhere. The Royal Commission on Coal Supplies in 1905 estimated that the great bulk of coal reserves in Fife and the Lothians contained seams that were considerably over two feet thick, whereas Lanarkshire was disadvantaged, with half of its reserves reckoned to be even thinner than two feet. From the 1880s in Fife and the 1890s in the Lothians, favourable investment opportunities were sought and found in the more intensive exploitation of the untapped, accessible reserves of the eastern coalfield. In the Fife and Clackmannan coalfield especially, the bulk of output was destined for export to the Baltic and northern Europe.[3]

In the pursuit of sustained profitability and improved efficiency, the structure of ownership and control within mining also underwent significant change from the 1880s onwards. Particularly in the west, much of mining remained in the hands of small masters and owners of one or two collieries, but everywhere there was an increasing trend towards the creation of larger firms by mergers and aggressive purchase of competitors. By 1913, the level of concentration of ownership, production and employment was such that the ten largest Scottish firms controlled almost half of the workforce in coal-mining. William Baird and Co. became the largest coal producer in the west, with over forty collieries before 1914; and the United Collieries, formed in 1898 with eight companies, swallowed up twenty-three others in 1902, to become a major force in the west-central coalfield. The Fife Coal Company became the principal operator in that county, after absorbing eight local companies between its formation in 1872 and 1909. As employers of colliery labour, William Baird and Co. and the Fife Coal Company were high in the top ten coal-mining firms within the United Kingdom. With 7,000 mineworkers, Baird was third in the UK league in 1894; and was in fourth position in 1913 with over 11,000 mineworkers. However, although the Fife Coal Company did not figure in the 1894 listing, by 1913 this firm had risen dramatically to become the largest colliery employer in Scotland, and the third-largest employer in the British mining industry, with 14,000 mineworkers.[4] On a smaller scale, the Lothian Coal Company occupied the dominant position across the Forth. By 1890, it was one of several coal companies in Midlothian to take over direct ownership and production from the traditional landed families. The Lothian Coal Company owned and worked the Newbattle collieries, and the opening of its Lady Victoria Pit at Newtongrange in the 1890s marked the beginning of highly profitable, mechanised, modern mining in the county.

Industrial relations in mining continued to be troubled, and coal-masters' associations were formally organised from the 1880s to counter the threat of trade unionism and the demands of the mineworkers on wages and conditions. The coal-owners and employers took a hard line against any limitation of their powers over recruitment, supply and payment of labour. Above all, they were concerned to keep labour costs as low as possible, and condemned the agitation for an eight-hour working day underground. This long-standing proposal of the miners, and its obvious potential for improving their health and wellbeing, was enacted in 1908 with Liberal and emerging Labour support in Parliament; but it became law in the teeth of furious opposition from the employers. They reacted in the same fashion to the minimum-wage campaign of the miners, and to the eventual legislation on this issue in 1912. By then, after a national miners' strike, the leaders of the Scottish coal-owners were united in their determination to reduce overall labour costs which, they claimed, had risen to around 70 per cent of total expenditure in their mining enterprises. In effect, intentions to cut costs meant cutting wages and increasing the rate of exploitation. Coal-owners were in business to reward themselves and their shareholders with lucrative dividends while putting the squeeze on workers who did most to create the wealth. The Fife Coal Company was among the most profitable companies in Scotland. Between 1882 and 1921, it paid out an average annual dividend of 20 per cent on ordinary shares. In some years, the returns to shareholders were as high as 50 per cent. Shareholders in the giant William Baird and Company enjoyed even higher average and special year returns on their investment. In 1913, representing those interests, Sir Adam Nimmo, director of the Fife Coal Company and chairman of the Scottish Coal Trade Conciliation Board, set the tone of implacable resistance to labour in the coalfields. With these telling words, he signalled the massive confrontations of the immediate post-war years and 1926: 'in any prolonged struggle between capital and labour, where both sides were fully organised, labour would have to give way, as the capitalist was more likely to hold out in a struggle in which starvation was the ultimate factor.'[5]

In the growing shale-mining sector, there was also an increasing concentration of production and ownership. The number of shale miners, all working on the stoop and room system, fluctuated around the 4,000 mark between 1900 and 1920. They worked for half a dozen companies who operated oil works in West Lothian and one in Fife.

By 1914, the largest company in shale mining was the Pumpherston Oil Company, employing over 1,300 pitmen at the main company town of Pumpherston and surrounding villages. The native shale-oil industry was under intense foreign competition, principally from America and Russia, and there was constant pressure to keep down costs of mining and production. Capital and labour costs in the Scottish sector were much higher than those of their large competitors, who enjoyed the natural advantage of crude oil gushing out of the ground and the economic benefits of petroleum production and refining on a big scale. As with coal, the importance of the native shale-oil industry was acknowledged and boosted in wartime conditions, but in 1919 all six firms were amalgamated into Scottish Oils Ltd as a subsidiary of the giant Anglo-Persian Oil Company. Thereafter, the familiar story of reducing costs to remain competitive in the world market was applied more ruthlessly than before to the jobs, wages and earnings of the shale miners.[6]

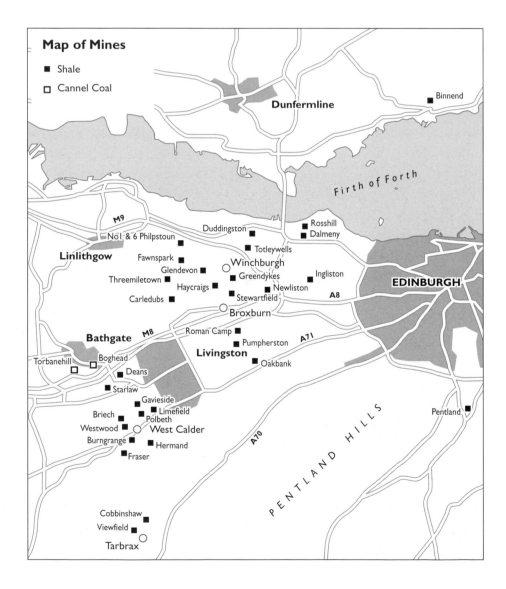

Plate 40.

Main locations of
shale mining
(*D Kerr,* Shale Oil)

The tiny lead-mining sector reached its heyday during 1914–18, but even then was producing only upwards of 5,000 tons a year during wartime. Thereafter, beaten by cheap lead imports, it went into irreversible decline. It failed at Tyndrum and elsewhere, although production at Leadhills and Wanlockhead lingered until 1928, by which time the enterprise was no longer commercially viable. For this remote industrial community, the loss of livelihood was devastating. However, this local experience was only a microcosm of the destructive closures and mass unemployment suffered by so many coal-mining communities across Scotland.

The Mineworkers:
Profiles and Personal Testimony, 1880–1920

The independent collier tradition survived into the twentieth century, but with great difficulty. Craft-conscious hewers and small family teams of face workers on the stoop and room and longwall sections found themselves under increasing pressure from unskilled labour, contracting and intensified and new ways of extracting and conveying coal. By the 1880s, longwall

Plate 41.

Workers at Glencrieff lead mine, Wanlockhead. (*Wanlockhead Museum Trust*)

mining had become more prevalent in all areas except for Clackmannan, Lanarkshire and the west. This trend continued, so that by 1914 longwall working had overtaken stoop and room everywhere as the principal operation in thin and thick seams alike. Moreover, coal-cutting machinery was firmly established, introducing the machine-run version of longwall. Fife and Clackmannan were pace-setters, followed by Mid and East Lothian. By 1914, 70 per cent of mines in the Fife district were using coal cutters, whereas the corresponding figure for Lanarkshire was 40 per cent. In Fife, the new technology was aimed at extracting the thick seams, whereas in Lanarkshire it was being used mostly to work the thinner seams. Stoop and room working and the hand and pick places on longwall survived most where machines were unsuitable due to geological faulting or were reckoned to be too expensive. By 1920, many Lanarkshire and Ayrshire mines retained the older hand-hewn system of coal extraction, and Ayrshire was still the least mechanised of all the coalfields.

Going down the pit to work for the first time was usually a frightening and traumatic experience for boys. As an old man, George Anderson recalled with vivid detail his first day when, straight from school and barely thirteen, he started with his father in 1907 at the Hill Pit, South Longrigg colliery, within the Larkhall-Dalserf district.

> In my youthful ignorance, I knew nothing of a miner's working life. Preliminary 'working' arrangements had been made by my folk. My working clothes were placed on a chair, separately, my boots beneath, similarly arranged as my father's. I ceased wearing shorts and was provided with 'long yins'. For working purposes I wore a belt around my waist instead of braces. Extra knee patches, acting as pad for protection, were sewn on. My jacket was an old one of my father's, with sleeves turned up at the wrist. Extra enlarged pockets were sewn on; they were necessary for carrying my piece box, tea flask, and oil flask for my lamp. I was also provided with a small oil lamp, which was hooked on a piece of leather sewn on to my cap at the front. This small lamp, shaped with a spout for holding the wick, gave light in front and downwards during work. The mine in which I started was of the 'naked light' class.
>
> I was already awake when my mother called me to get me up for work at 6 a.m. The walk to the pit head was only a short distance from home. Along with my father, I joined a group of miners gathered at the point near the pit shaft, and waited for the cage

which would take us down the shaft. My previous excitement and anticipation was now replaced by doubts and something akin to fear. However, finding myself suddenly 'in turn' with the cage already waiting for me, I had no further time to think but act and walk automatically into it. My father, holding me by the arm, gave me the courage to walk into the cage. Standing close by his side, clinging to his jacket with one hand, I followed the example of the others in the cage by stretching upwards with my other hand in an attempt to catch hold of the cross-bar. While suitably placed for adults, it was, owing to my diminutive height, as much as I could do to reach it without a struggle. Moments after entering the cage, I felt myself hurtling downwards through the shaft in complete darkness. I struggled desperately to get a tighter hold on the cross-bar despite the water running down my jacket sleeve. I dare not slacken my grip, for there was at that time no safety gate to protect me from the open shaft. With my other hand I clutched tightly to my father's jacket. He, in turn, held me close to him and this gave me a greater feeling of safety. A strange uncomfortable feeling came into the pit of my stomach as the cage continued its hurtling descent. This sensation remained with me on many a later journey.

The cage with its almost silent running, sliding effect on the wooden guides sped downwards in the darkness. How I longed for it to stop. Eventually, much to my relief, the cage slowed gently to a halt at the pit bottom.[7]

Frank Hillan, who started in 1914 as a drawer in the upper Lanarkshire coalfield, driving a pony and full hutches from the coalface to a main road, echoed these sentiments:

I'll never forget the first mornin when I stairtit at the pit at Coalburn. When I went doon in the cage to take us underground, it drapt so fast that I thocht everything had left me. I swore that if I ever got back up the pit, I would never go doon again – but the next mornin' I had to. It had been a big day in oor lives when we said goodbye to attending school. We went tae the pit dressed in auld claes – a pair o' trousers and a jacket. We had bits o' leather sewn onto the front o' oor bunnets tae hold oor oil lamps as it was in the days before the introduction o' carbide lamps that provided better illumination. Doon a pit wis a terrifying experience for a boy like me who wasna even fourteen. I remember how the boys startled at every noise. Even a drop o' water falling on a stane could be quite frightening.[8]

Now Anderson continues:

> In the darkness, my father led me out of the cage on to a floor of steel plates, and over to two open-flame, large-sized lamps. Along with others from the cage, I was able to light my own lamp. After a few minutes' delay to allow me to adapt to the darkness, I was then ready for the next move.
>
> The main roadway at the pit bottom had plenty of head room. The whole area had the atmosphere of an underground railway. I found things were very different after leaving the main roadway. The feeder roadways were lower and you had to bend when walking and the air became warmer. As the feeder roads were narrower, you had to squeeze between passing or stationary hutches on the track and the walls of the roadway. The journey extended for about a mile and the further we travelled the more the temperature increased. On account

Plate 42.

Going down in an open cage, without protection or safety. (*North Lanarkshire Council*)

of this and the necessity of dodging rollers and steel ropes, the journey soon began to tire me. Ultimately, we arrived at a place known as 'the fireman's station', where a number of miners sat waiting. After my father had received word from the fireman that all was right, we branched into a roadway leading directly into our working place. I was sweating, had bumped my head a few times, and was wondering when all this travelling would cease. I wanted to sit down and rest, but that was not permissible since we were nearing the coal face, and since we needed to have some coal ready for starting the drawer, I had to reluctantly keep walking. At that time, I did not appreciate that this walk would be a daily routine prior to starting work at the coal face.

Near the coal face I took off my extra clothing, rolled up my sleeves and was ready to start work. Walking towards the working space, I came to a solid wall of stone stretching the height and width of the roadway, but only reaching to about thirty inches from the floor. I knelt down on the pavement and peering beneath the rock I saw the shining black face of a wall of coal for the first time in my life. I then crawled in below to get a closer look at this lovely sight. As I sat looking around me, having asked questions of my father, I found that his working stretched along this coal face for about fifteen feet on either side of the roadhead. Looking along to my left I saw the light of another miner's lamp reflecting in a pool of water in my father's place. What amazed me was that my father had to work in this pool of water under such a low roof, it being only thirty inches high.

My father showed me how to shovel the coal from the face to the roadhead. Getting down on my knees, I reached out for the coal, trying to avoid the props and flung it over to the roadhead for filling into the hutches. Lacking the knack, I suffered cuts and bruises and the floor was unmercifully hard on my knees. I changed to a sitting position in the thirty inches space but this, too, soon became painful and tiring. My father, seeing my exhaustion, took over and told me to rest for a while. I realised that I was a boy trying to do a man's job.

A rumbling noise, growing louder and louder, signified that the drawer had arrived with an empty hutch. Later, the delay in the drawer's arrival indicated to the miner that it was corning time, otherwise lunch. My father examined a place a few yards from the coal face and decided I was to sit there. My lunch, three slices of bread with cheese, warm tea from my tin flask, I really enjoyed. Being so hungry, the surroundings did not bother me sitting down on the floor of stones and dodging water dripping from the roof, something which would be experienced many times in my future years as a miner. After eating, a miner from each working place beside ours joined our company in conversation. Simultaneously, the two boys from their working places joined me. We were being kept outwith the elders' conversation, being considered too young to take part.

The allotted half hour seemed to go in like lightning, as the sound of the hutches was heard once more. Hutches received at the coal face had to be filled and ready for the drawer to uplift on restart time. This, in effect, meant that only twenty minutes were allowed. However, miners being on tonnage accepted this shortened lunch time.

Filling the hutches, putting in the pins, and going below the laigh shovelling coal from my father at every available opportunity for the arrival of the drawer kept me busy all the time. Ultimately, the last hutch was filled and we gathered up our tools to a safe place, put on our jackets again and made our way back towards the pit bottom.

Later that night in bed my back and shoulders ached. I had cuts and bruises on my hands and arms and I was so tired and sore that I could not sleep and had to lie on my back to ease the pains in my shoulders.

George Anderson was born into Scots mining stock in Cambuslang, Lanarkshire. His father and grandfather were colliers, and he took the typical path into pit work as a boy beside his father at the age of twelve years and ten months, although the official school leaving age was then thirteen. As George explained,

> I had no say in the matter: the choice was made by my parents. An exemption certificate from the local school board had to be obtained. The basis of the claim for exemption was financial. I was the eldest of a family of seven and the extra money earned by me would help both myself and the remaining members of the family at school.

His father was sole breadwinner and a full week's work as a hewer was not guaranteed. A bright boy at school in the mining village of Netherburn, three miles from Larkhall, George could have expected to progress through the Supplementary Class Examination at thirteen and perhaps then gain entry to secondary education. However, as was the case in impoverished mining households, this chance was lost and he was obliged to make a vital contribution, however small, to the family wage. A contemporary, Will Blackwood, from Douglas Water, born in 1894, left school early for the same reasons: 'Because I was the eldest of a large faimily, I got an exemption frae attending the schule, and on 26 June, 1907, I stairtit. I got handed an oil lamp to attach to ma bonnet and went doon the pit wi' ma faither.'[9] Jim Hastings, brought up in Glespin, on the Douglas coalfield, was also a victim of family hardship and parental pride, leaving school early for pony driving at the surface: 'I left the schule when I was thirteen. I got an exemption, as we were badly aff in the hoose. Ma faither wis aff wurk wi' pleurisy and pneumonia. Ma parents wouldna ask for any money from the parish.'[10]

There was yet no formal system of apprenticeship for boys who aspired to learn the skills of the hewer and all-round face work but, as Robert Smillie, the Lanarkshire miners' leader explained, 'it is largely the practice in Scotland for the lads to go to the working face with their father when they leave school … we call a skilled Scotch miner a person who has been brought up at the colliery from 13 or 14 years of age.'

In similar fashion, John Weir, from the Fife and Kinross Miners' Association, observed, 'at our collieries the young people go in with their fathers and gradually learn to become miners.'[11] If the father and son pairing for the immediate coalface was not possible, the next best option was to keep recruitment and co-working within the family-based unit, so that boys could get a start with a brother, an uncle or in rarer circumstances, with a grandfather. Failing that, attempts were made to fix up the boy with a family friend or a trusted neighbouring collier. There was an expectation that such miners would be most likely to take responsibility for their own boys and family, treat them fairly and train them up properly as safety-conscious, skilled colliers.

Where traditional practice was sanctioned by the trade union and accepted by managers, on each shift the boy face worker was entitled to a certain number of hutches, which had agreed value as part of the full man's wage. In this regard, George Anderson's account is consistent with that of Smillie for the 1890s and early 1900s in Larkhall district and beyond. Smillie talked of young boys of fourteen being worth half a 'turn' or 'ben'; and three-quarters up to sixteen. According to Anderson, 'every boy working at the coal face, whether with his father or a stranger, was on "half ben". He was only entitled to half share of hutches per man. If a miner got six hutches to fill, a boy would receive three, which would mean a total of nine hutches altogether. With two miners working together, they would receive payment for twelve hutches.' In this way, without stoppages to production, a hewer and assistant, whether man or boy, had the opportunity to fill the available quota of hutches and make up a reasonable wage if the colliers considered that the current tonnage rates were viable. According to Anderson, it was up to the discretion of the hewer how much to pay the boy. As indicated above, his first and main job as trainee face worker was to shovel the coal from underneath the worked seam out to the roadway and then into the waiting hutch to await the arrival of the drawer. 'As usual, circumstances and the boy's ability decided. One must have his hutch full, ready to keep his ben and receive the empty one. Failure to do so meant losing his turn for the hutch.' One of the first essential tasks was to learn the knack of filling a hutch speedily without overloading and at the same time achieving a 'good weight'. 'To achieve the best results, large pieces of coal could be arranged along the top of the hutch to obtain the maximum weight but arranged in such a way to avoid "roofing" during transport.' If the hutch load made contact with the roof, dislodging coal, which had to be retrieved from the roadway, it posed problems for the drawer, who could get his revenge on the face worker by further delaying or withholding his hutch.

Besides shovelling and filling George learned all the coalface jobs – undercutting and holing with pick, roof-testing, shot-firing and blasting and propping – before he gained his own 'place' at sixteen. Of all the tasks, he emphasised that drawing a rake of heavy hutches on the sloping railed roads was not a job for boys of fourteen or fifteen:

This job was only suitable for young, active and strong men. They had to be able to push hutches up steep gradients and control full ones coming down. On steep gradients the wheels of the hutch were completely stopped and the hutches allowed to slide on the rails while the drawer held on to control the speed. This control was sometimes applied by the drawer walking in front to slow it down with his back. This was a dangerous job. A slip of the foot could have fatal results. The backs of the drawers were often skinned and bruised.

Shovelling, filling and drawing was a brutal experience for a growing fourteen-year-old, as Archie Henderson discovered during his first weeks in a Shotts pit:

My job was to shovel coal to a roadhead and fill it into hutches. Much of this work was done on my knees, and they were very sore and bled. I learned that I ought to have had padding. My mother provided for me. War came in August 1914, and many of the young strong miners joined the army. It was then that boys were asked to do men's work. My feet had grown and my boots hurt cruelly but I did not want my mother to buy new ones. Shoving hutches, the pain was severe, and I had occasion to weep with agony. Not long afterwards something happened to my legs, the muscles were seized with such pain I could scarcely walk. Unable to walk I set out for home. The journey took me hours. On the road to the bottom I met the gaffer and I described my condition. He listened and said, 'Och, man, ye've jist got growing pains, ye'll be better the morn.'

He then shifted to another pit to assist a new hewer, only to find his problems continuing, and the pace of heavy work intensified. Here, 'the working was very wet, and at night I felt as if my legs were being stabbed with daggers.'[12]

The experience of John McArthur as a boy collier on longwall working at Muiredge Colliery, Buckhaven, provides additional clarification from the Fife coalfield. He started with the Wemyss Coal Company in August 1913, in accordance with the Coal Mines Act of 1911, which made fourteen the

Plate 43.

Miners working at narrow seam in a Lanarkshire colliery. (*North Lanarkshire Council*)

minimum age for employment underground. He began as filler and drawer with a miner who was friendly with his parents. He describes the work, as follows:

> Miners at that time usually worked in pairs, a man and a youth … the face man, or hand-got man, won the coal, secured the roof, built the wooden pillars to form the roadway, and laid rails along which were moved the wooden hutches, each holding nine, ten, or eleven hundredweights of coal. … the youth filled and shovelled the coal won by the face man. After pushing the hutch to the 'lie', or wheel-brae, the youth, by means of a detachable steel chain wound round a pulley, sent the full hutch down the incline towards the pit shaft, and by gravitation the movement of the full hutch pulled up an empty one for filling. The process went on throughout the shift. He also helped the face man with the work at the face, and gradually served his apprenticeship in this way. The filler and drawer was usually paid a day wage, whereas the face man was normally a piece-worker. I worked for three years as filler and drawer before becoming a face man at the age of seventeen.
>
> As a boy of fourteen, I thought I was lucky to get the princely sum of 2s. 9d (about fourteen pence) a shift. That was the standard rate for a boy of my age. The hours of work were eight a day plus winding time. That is, the miners began to be lowered down the pit half-an-hour before the shift actually began, so that they would be at their place of work a full eight hours.[13]

The regulation hours of work were a consequence of the Eight Hours Act of 1908 and his 'standard' wage was in line with recently won minimum wage legislation.

Like Anderson, he proudly praised the skill and versatility of his collier father:

> My father was one of the old school of mining. A highly skilled picksman, it was a treat to watch him working. As a result of training, he was ambidextrous, could shovel with either hand, and use the pick lying first on one shoulder then on the other. The steady beat-beat of that hand-got pick was a work of art. Old miners like my father were carefully trained men that performed the complete operation of a miner. They undercut the coal, kept enlarging the cutting until they would get four or five feet undercut. Then they

would use a hand-boring machine, bore holes, make up their own explosives and blow the coal down, then form the roadway, help fill off the coal, and secure the roof.

For a face man my father used a wide variety of tools. He used light holding picks, some long in the grain, some shorter. He had heavier picks for cutting the coal and breaking it up, heavier picks still to deal with stone work. For coal that was easy to bore he had a fast borer, for difficult or hard coal a slower borer. He also had a cleaner for cleaning borings out of a hole, a stemmer for stemming the shot hole, a heavy hammer or mash, a mash-axe, wedges, splinters, and so on. He had almost a hutchful of tools.

McArthur added that this expertise and variety of tools contrasted with those of a stripper – a hand miner assisting on a machine-cut face – who could not be classed as a skilled man. The stripper relied only on 'a shovel, a pick, sometimes a pick with a mash end, and a mash', and was not responsible for taking care of the remaining condition and safety of the coalface.

The spirit of the independent collier tradition was very much alive in the concern of skilled picksmen like McArthur to make their own work bargains with the managers, where possible, and to stay clear of the further exploitation that arose from working in a squad for a contractor. In their case, two families, including the McArthurs, teamed up to take their own contract to develop a new section at the local Wellsgreen Colliery. For such a complex and highly skilled job, only trusted and experienced men were required, and they were employed directly by the managers.

The collier's typical strong sense of craft pride and desire to have as much control as possible over the work process was also evident in McArthur Senior's initial resistance to involvement in machine-cutting. As his son relates, the occasion arose under pressure of wartime production in late 1914. The family team were still opening up the section at Wellsgreen, and it was not yet in a position to produce a large amount of coal. The company decided to postpone further development work, and to concentrate instead on working high-output sections for the war effort. The McArthurs, like so many others, were sent to a section where undercutting was done by machine, and the work was organised by a contractor. However, their work role as an adjunct to the machine lasted for only a short period:

This section was only about three feet high, in contrast to the section we had always worked in, which was four-and-half to five feet high and where the coal was hand-got by the pick. The coal-cutting machine was a revolving disc with picks attached that did the under-cutting. The holes had to be bored by a hole borer, the roadway was made by brushers on the opposite shift. The job of the miner was to blow down this coal with explosives and fill it off. Because the section was so low we had to work on our knees. My father was unaccustomed to this form of work.

The story of how he, his father and brothers, left this work is both instructive and humorous:

> I was filling and drawing my father's hutches, with about ten hundredweight in each hutch, transporting it fifty to sixty yards down to the main junction, when Muir (the contractor) said to me: 'See and tell your father to lift his pugs.'
>
> I did not know what that meant but I duly repeated what the contractor had said, 'Dad, you've got to lift your pugs.'
>
> 'What's that?' my father said, 'What the hell's pugs?'
>
> 'I don't know', I said, 'but you've got to lift them.'
>
> 'Son', he said, 'that's the best thing you've said the day. Here ye are, lift them and put them in the tub.'
>
> Pugs in fact meant coal the machine had failed to cut, but it was a new word to my father, who had never worked in a machine-cut section before. He was fed up working in unfamiliar conditions, so when I reported that he had to 'lift them' he thought it meant he had to lift his graith, or tools – that is, to pack up and leave the pit altogether. The contractor had not meant that at all, but my father had no regrets at leaving.
>
> This complete change in methods of work had a serious effect upon picksmen miners like my father and brothers. They had to look for work away from machine mining if possible.

McArthur and sons had no trouble in finding a return to hand hewing methods at Muiredge and, in any case, there was plenty of demand for their labour, as Fife had the highest levels of enlistment into the armed forces within the Scottish coalfield.

John McArthur, and colliers like him, who had learned from his father in the traditional manner how to become a multi-skilled face man, continued to prefer hand hewing and the practice of all-round tasks, which ensured his status as the main production worker, and conferred some job satisfaction. Under a machine-cutting, contracting regime, the 'complete collier' was an endangered species. A new, strict division of labour prevailed, in which the machine-man did the undercutting, a borer then drilled the coal and a shot-firer charged the bored holes with explosives before bringing down the coal. So often, the experienced collier, who had formerly carried out all these operations, was reduced to little more than a shoveller and filler of coal in a machine squad. In areas where the new cutting technology was being extensively introduced at the expense of traditional methods of extraction, the balance of the work process shifted towards de-skilling. For each machine man, up to fifteen or twenty fillers could be required on a shift, many of whom were usually raw recruits or otherwise inexperienced miners. However, at this unskilled end of operations, such squads would also include former skilled hewers who had little or no choice in taking on low-status tasks.

By the early 1900s, it was also clear that machine cutting had begun to set up a new hierarchy of skilled men underground – of machine men operating electronically driven cutters; electricians and mechanics; and other trained men for face, roof, road and rail maintenance. In this changing work environment, the experienced, versatile collier was well aware of the threat that machine mining posed to traditional skills and to his standing as superior face worker. Moreover, as machine mining became more commonplace, and the depressed state of the mining industry took hold during the inter-war years, fewer opportunities would exist for coal miners to move easily to pits where hand hewing was still practised and rewarded with a living wage.

Miners had large families, and they continued to provide new generations of young males for the pits. Clearly though, from the 1890s until the early 1920s, this last great expansion of coal-mining required a huge labour force, which could not be supplied from this source alone. More than ever before, recruits had to come largely from outside the industry and the existing mining communities. Whereas hewers usually continued to come from mining families, and to be trained up by them, raw and unskilled outsiders were increasingly being introduced into underground work, hired and directed by contractors who tended to drive down wages and increase the workload.

Vulnerable Irish immigrants remained an important source of labour for coal- and shale-mining from the 1880s. The census for 1911 records 12,000 Irish-born males as mineworkers in Scotland, consisting mostly of the latest generation of recruits into mining. The skilled Scots colliers and their spokesmen like Keir Hardie kept up a long-standing protest about such men being used in the thick seams and the best places: 'Nothing angers the miner so much as to find a fellow working at stoop, where the requisites are a big

Plate 44.
Overleaf. Lithuanian miners. (*North Lanarkshire Council*)

shovel, a strong back and a weak brain, said fellow having been busy a few weeks before in a peat bog or tattie field and who is now producing coal enough for a man and a half.'[14]

In the same vein, they objected again to 'grown-up Irishmen coming into the pits as drawers, and in a few months being allowed to have places of their own. It does themselves no good and the trained colliers much harm. ... just now any young stirk of a fellow can come into a pit and call himself a collier; but if all the men in the pit were such, then the pit would go to wreck and ruin.'[15]

The objector was referring not only to the intrusion into mining of the untrained Irishman but also, in years of depressed trade, short-time working and unemployment, to the entry of labourers and semi-skilled men from the likes of the iron and steel sector, the building trades and from farm service.

From the 1880s until the war years, a significant new source of male immigrant labour was employed in mining. Like the Irish before them, incoming Lithuanians (loosely described as 'Poles' by their Scottish hosts) were the subject of considerable concern and resentment among native colliers. Lithuanians were an oppressed nation within the Russian empire. The 8,000 or so Lithuanians who had arrived in Scotland by 1914 were predominantly Catholic, some were political refugees fleeing from persecution under Tsarist rule, most were poverty-stricken and seeking a better life elsewhere. The bulk of this immigrant group were single men, many from an agricultural and rural background, and used to low wages and poor living conditions.

The first few arrivals in the 1880s were recruited by agents for large iron and steel companies in Ayrshire and Lanarkshire. They were enticed into the black country of Scotland with the promise of work and housing. At Glengarnock, they were employed by Merry and Cunningham as unskilled blast furnacemen, lured into dirt-cheap labour in atrocious working conditions. Some then went into mining in the same area amidst claims that, in their ignorance and their acceptance of poor conditions, they were being used as non-union reserve labour to break strikes and to keep down wage and tonnage rates. Moreover, as the foreigners were unskilled labour, without knowledge of the English language, and unfamiliar with the hazards of the mining environment, native miners claimed that they were a danger to themselves and their fellow-workers. All these allegations against Lithuanian labour in the mines can be tested against available evidence, mainly from Lanarkshire, where their presence was concentrated.

It seems that Lithuanians first came to work in the North Lanarkshire coal mines in the early 1890s. By 1901 there were more than 1,000

Lithuanian miners in Scotland, rising to 2,600 in 1911. They worked and settled for the most part in the mining and industrial communities of north Lanarkshire, in particular around Bellshill and Motherwell. In 1908, about 200 Lithuanians were employed as miners by the Lothian Coal Company, residing in Newtongrange and other nearby company villages. Others also found work in coal- and shale-mining in West Lothian.

Until the early 1900s at least, there is evidence that Lithuanian labour was used to cut wage rates and to break strikes. As with other raw labour, employers undoubtedly took advantage of the immigrants' poor bargaining status to help reduce their labour costs. For instance at Thankerton, Tannochside and Neilsland, in 1902–3, Lithuanian or Polish miners were brought in, accompanied by a reduction in tonnage rates. The last recorded instance of this nature was in 1909, when it was reported that Poles had been started at Watson's, Motherwell, at rates below standard.

There are also several known instances in the early 1900s, notably at some Hamilton, Motherwell and Glasgow collieries, where Lithuanians were brought in to weaken trade union solidarity and to break strikes. As one would expect, the expressed views of the mining employers were usually complimentary about the hard-working, reliable and submissive foreign workers. Contrary to the viewpoint and evidence of the miners' union, the employers were also adamant that the foreign workers did not present an additional danger to safety. They claimed that adequate provision was made to instruct them in their duties, including translations of mining regulations in their own language. This point was countered by the workers themselves, who were angered and humiliated when faced by written instructions in Russian – the language of their former oppressors. Referring to 'Polish labour' in the Scottish mines, an articulate voice from the native mining community stated objections on safety grounds:

> The majority of these have no knowledge of either English or mining.
> Several mining firms have tried to overcome the language difficulty
> by having rules applicable to mining printed in the language of the
> alien. This made no difference because only a very small percentage
> of them could read the language they were able to converse in.
> Meantime, the employment of alien labour is a considerable source of
> danger to the other mineworkers.[16]

It is difficult to ascertain just what proportion of foreign mineworkers were involved in accidents, or whether they were more prone to injure themselves and others around them. The record is unclear as many Lithuanians adopted

other names, including those of fellow Scottish workers, and thus entries in accident and colliery books are not easily traceable to this minority group of workers. Union officials could cite plenty of instances of Lithuanian and Polish workers being involved in accidents, including the awful case of a man who was decapitated when he did not understand the engineman's warning to keep his head in as the open cage descended the shaft. There are other stories of failing to observe warnings about the presence of gas and of removing glass from safety lamps to improve the amount of light. Yet, such instances could not be used to fix the principal blame for neglect of safety on foolish, ignorant or unskilled, inexperienced workers, whether foreign or native ones. Instead, all safety-conscious miners consistently pointed out that existing safety regulations were either inadequate or were not properly enforced and observed by managers and officials, even by hard-pressed firemen who had first-line responsibility for such matters underground. In this period, from the trade union side, leaders such as Keir Hardie, Robert Smillie and Robert and William Small constantly made representations to official and government enquiries, voicing their concern about the combined inadequacies of underground safety provision and the employment of unskilled men of all kinds and origins. In making their case, they were especially critical of unscrupulous contractors, who employed unskilled and inexperienced men and drove them carelessly, and alleged that this group of bosses was among the worst offenders against safety regulations and procedures.

Nevertheless, at this time, the standpoint of the employers was generally endorsed by the 'official view' of inspectors and of Government that unskilled and foreign workers did not constitute an extra danger to themselves or to others underground. From another perspective, we shall see later that Lithuanian miners were not only good workers but also became strong trade unionists and were committed to the labour movement.[17]

Safety Matters Below Ground

Considering safety issues more generally, it is not surprising to find reports and descriptions of a catalogue of accidents and near misses in the personal testimony of miners during this period. For example, George Anderson recalled the frequency of eye injuries, particularly to the hewer, from particles of coal. One of his early jobs was to attend to his father,

> who had got a small piece of coal in his eye and came to me with one hand covering his injured eye and holding out with the other the pricker for his lamp wick, telling me to remove the particle. It was

my first 'first aid' adventure and a lesson for many future occasions. Because of the pain created, after coming home, a common practice was to fill a basin with cold water and wash the face with the eyes open to reduce the inflammation. Where eyes were permanently damaged, in some cases gauze goggles were used, but mine darkness was a handicap with them on.[18]

It was an eye injury that finished the career of his namesake John Anderson as a face worker in December 1913, while picking hard splint coal in a longwall place at Gilbertfield pit, Cambuslang:

A burst of coal struck me on the face and knocked me over into the waste. After I regained consciousness I informed my drawer of my accident. He told me my right eye was in a bad state and advised me to go home. I went home and saw the doctor that night. He dressed my eye and bandaged it, and ordered me to take a week or two of rest. As my eye did not improve he sent me to the eye infirmary. After treatment there I was told that the sight was gone as the blow had separated the optical nerve from the eye. The employers would not admit liability, but six months later the Union fought my case in the sheriff court at Hamilton, and they won it for me. I got full compensation for six months, and was then put on light work on the surface with part compensation.

John Anderson was angry and bitter about the needless dangers he and others had to suffer and about the arbitrary behaviour of officialdom above and below ground towards the everyday victims of that harsh and hostile environment. He was particularly scathing about the realities of mines' inspection in his own time as a working miner. In his considered view, mines' inspection was 'only a farce', as they 'were so few that they could only report serious accidents. I may say that as a miner of thirty-two years' experience up to 1913, I saw only four inspectors in all that time, when they came to report on an accident or explosion.' He added that miners were afraid to report hazards such as poor ventilation to inspectors for fear of victimisation and dismissal, and that only when the union was sufficiently strong was it possible to make any meaningful representations to the management on safety and compensation issues. He recounted many hazards and incidents from his own working life in thirty-nine pits, mainly in the Cambuslang area. The following instances are related below, as they were probably typical of the experiences endured by his mining contemporaries:

I was working on a level in the virgin coal in no.2 pit with a mate on the afternoon shift. After preparing to start work, I walked up to the road-head, which was high, and thinking that the place was safe, as there was a boy holding a hutch at the time, I stood up straight. On doing so, my naked lamp, which hung on my bonnet, ignited a pocket of gas, which went off with a roar and blast. I threw myself on the pavement as I felt the furnace of gas going over my body. When the gas had exhausted itself I stood up to find my back excessively hot. Putting round my hand to find the cause for this, I discovered that my brand new flannel shirt was on fire – a shirt bought the previous day at a cost of 6s, which sum was a lot of money to my wife in those days. Fortunately, that was the only damage done; the rest of my mates only suffered from slight shock. Later, a new section was being opened in the soft coal. The working was very wet, and as the contractor would not give us oilskins, we were a sorry sight to see when our shift was finished. I worked with only my trousers and boots on.

Some weeks later, owing to an extension of the workings, the air became very bad for the want of a communication. The powder smoke and lamp reek filled up the vacuum where we were working, which made it difficult to breathe. We requested the oversman to put in a fan. As this meant expense and a boy to be paid to work it he told us to carry on as the communication would soon be through. One day I suggested to my workmates that we should buy two or three candles and bring them out to the pit. Next day, prior to the daily visit of the oversman, we put away our lamps and lit the candles. When the oversman saw us working away in the fitful gleam of the candle light he was very angry and wanted to know what was the idea. As no-one answered him, I had to tell him that this was how the miners, centuries ago, worked when the art of mining was in its infancy and modern artificial ventilation unknown.

Next morning there was a fan installed in our place. He asked the contractor to sack me. To his honour he said he found no fault with me as a workman, but later the oversman put my mate and me into an abnormal place where the coal was much thinner, and because we could not load the same as formerly he stopped the place and we were thrown out of work.

We went to Gilbertfield pit where, as the coal was extra hard there were a lot of explosives used. The air became so bad that it was nearly impossible to see our neighbour or load the hutch. Often at the end of the shift the temples were like to burst, the balls of the eyes were red, and with a groggy feeling about you, you could scarcely walk steady; when you reached the surface you were about to collapse.

A few months after, this, my left hand began to swell and my fingers became so painful that I could not work. I went to the pit doctor who said I had got blood poisoning off the coal as he had similar cases in the same pit. After two weeks of terrible pain he cut my two fingers with a lance and told me to poultice them. A week later he ordered me back to work. He would not sign my line for compensation. As the wounds of my fingers were lying open my wife had to dress them night and morning, but my work and the coal dust would not allow them to heal up. A few days later the oversman told me if I did not go to the office and see the insurance doctor and sign off, my place would be stopped. The insurance doctor asked me how I was able to work with such sore fingers. I told him the pit doctor made me as he stopped my compensation. While at work the pain was excruciating. He told me to take another two weeks. I claimed money for loss of earnings. He told me to get rid of the pit doctor as they punish men for cruelty to animals, such as open sores in horses, let alone work in a coal mine with open wounds in one's fingers.[19]

Crude surgery to amputate a thumb caught in the cog of an underground conveyor was a traumatic experience for James Darling as a teenager during the 1920s. He describes his treatment at the doctor's surgery in Eskbank:

Ma father went wi' me. Dr. Mackenzie brought a basin. Ah put ma hand – this stump – intae this hot water and Lysol. It wis murder! And he came back in wi' a pair o' scissors. Ah seen them, they were big scissors, and they had a grip on them. Ah never thought for a minute he was goin' tae start on the thumb wi' this. He says, 'Turn your head away.' So ah turned ma head away an' held ma hand up. He took a haud o' it wi' the scissors – cluck! Phew! Right through the bone there! Ma knees struck me under the chin, wi' the nerves, ye know. And then ma auld man went mad. 'Cry yersel a doctor? Ah widnae ha' done that wi' an animal.' So he stitched the thumb up after that. He brocht it back and stitched it up. Ah never got any anaesthetic or nothin'. But ah'll tell ye something. That wis nothing compared tae what it did tae the nerves. Because ma thumb healed up but the nerves never healed up. Ah should have been taken tae hospital and put under an anaesthetic tae get that done.[20]

Apart from such accidents and incidents, the usual catalogue of disasters involving large loss of life continued to dominate the headlines during this period of coal-mining. Only a stone's throw from the scene of the Blantyre disaster, at Udston in 1887, explosions ripped through the dry, dusty and gassy pit, killing seventy-three workers in the second highest fatal incident in Scottish mining history. An initial gas explosion had set off other gas and dust explosions and, following this particular disaster, it became a legal requirement to scatter limestone dust throughout underground workings to keep down coal dust levels. Ventilation failure also caused the deaths of sixty-three men and boys caught up in the smoke and fumes from a boiler fire at Mauricewood colliery, Penicuik, in 1889. Between 1893 and 1901, an aggregate total of over fifty lives were lost in six other disasters, involving explosion, fire, flood and inrush of earth and moss from above. Later, at Stanrigg, Lanarkshire, in 1918, twenty-two lives were claimed by invasion of earth. In 1913, at Cadder, near Bishopbriggs, twenty-two workers were killed by suffocation in an underground fire. Untrained rescue teams of local miners battled bravely for days in hellish conditions to save lives, claim bodies and put out the fire. The Lanarkshire Coal Masters' Association was criticised severely in the subsequent inquiry for failing to abide by the safety provision of the Coal Mines Act of 1911 to set up a dedicated rescue brigade and station in their locality. Further, the owners were accused of risking the lives of ordinary miners and limiting their chances of saving trapped men and boys in such desperate rescue attempts.

Despite precautions, regulations and trade union agitation to reduce the dangers of work in coal-mining, the constant toll of death and injury continued to beset the industry, so that by 1914 the horrifying statistics tell us that one miner was killed in Scotland every six hours, and one was severely injured every two hours, mostly from roof falls.

The safety record in shale-mining during this period of its greatest expansion was equally as bad as in coal-mining. The actual number of accidents and injuries is not known, but many incidents involving injury prior to implementation of the Workmen's Compensation Act in the late 1890s went unrecorded or were under-reported. Between 1881 and 1914 there were 245 deaths arising directly from accidents in shale mines. Roof falls and runaway hutches in the extreme inclined roads headed the list of causes, followed by firedamp explosions and use of explosives in blasting. The official record of 434 injuries in 1911 makes it clear that one in ten shale miners sustained a serious injury that year. Beyond that recorded figure, there is no accounting for the toll of disability and eventual death arising from dust poisoning occasioned by exposure to the smoke and fume particles from explosives and blasted rock.

Drilling shot holes and blasting with explosives were more frequent operations in shale-mining than in coal-mining, and incidents with explosives were more common among this sector of workers. Problems with igniting a number of shots at the same time, and with misfiring, added to such dangers.

Shale miners were also more susceptible to firedamp explosions. From the 1880s, reliance on the stoop and room method left large areas of unventilated old workings, which were not sealed off. Usually, as the roof settled after mining operations gas would seep through the fissures and anyone with a naked lamp straying into those waste areas to evacuate their bowels or look for loose shale could cause an explosion. Although safety provisions in the Coal Mines Act of 1911 also applied to shale mines, the use of naked lamps was not completely banned. This exemption, won by the employers, was apparently due to difficulties encountered by working with the reduced light of safety lamps in the thick, pervasive atmosphere of gunpowder reek and dust that was so prevalent in shale-mining operations. Shale miners were spared the introduction of machine cutters into their working places, owing

Plate 45.

Shale miners, with open-flame lamps.

(*West Lothian Council*)

mainly to the obtuse hardness and shape of shale. However, as shale miners were already exposed to extreme noise and dust, it is doubtful whether machine cutting and its associated blasting would have made much added difference to their levels of danger and discomfort at work.[21]

Female Pit-head Workers: Profiles and Testimony

After their exclusion from pit work in 1842, women and girls were not prohibited from working above ground in collieries. Even so, during the Victorian era and beyond, the debate continued about the suitability of such work for daughters, future mothers and wives within mining communities. Without bothering to consult the working-class women and girls, there was a general consensus in middle-class society and among males outside and within the mining community that the proper role of wives was in the home. Their daughters were expected either to assume domestic duties alongside the other womenfolk or find respectable employment away from the coarse working environment and rough language of the pit.

However, the employers of pit-head lasses thought and acted differently, as they had a source of cheap and reliable labour for various unskilled jobs, often preferring to engage them instead of men and boys. As for the women and girls who took up such work, they were usually responding to the hard realities of the economic circumstances they and their families faced, by finding a ready source of waged work, however small, and making a contribution to family earnings.

In the 1880s, there were fewer than 700 females at pit-head work in Scotland, over 600 of them in Fife, where there was a continuing tradition of female labour in and around pits. Until this time, only a trickle were employed in the west, where there was no such tradition, and where male workers had always sought to assert their rights to all pit work. The number of pit-head lasses increased towards the end of the century and into the next, so that the 2,300 workers in 1911 were found mainly in Fife and Lanarkshire, the others in Stirlingshire and West Lothian. Moreover, nearly all were under the age of twenty-five and unmarried, in line with the established view that waged work ended after marriage. Their increasing presence at the pit head reflected the sheer demand for labour brought about by the expansion of production from deep mining and the need to inspect, sort and clean the various sizes and types of coal according to particular market demand. As the vital work of screening and hand picking became more widespread, so did the idea become more accepted that women and girls could perform these tasks, although the miners were divided on this issue. According to regulations in 1887, pit-head women were prohibited from undertaking the heavy work of loading, tipping and dragging wagons and hutches, but this ruling was not always observed. Before 1877, women and girls as young as ten were employed on the surface up to twelve hours a shift, including night work. Thereafter, their working hours were confined to daytime, including Saturday half days, and the minimum age for girls was raised to twelve. The law on working age then changed in line with the school-leaving age to thirteen and then fourteen.[22]

To illustrate the nature of the tasks and the attitudes of the women and girls towards their work and lives, three personal accounts are taken from different parts of the Scottish coalfield in the early twentieth century.

Maggie Spence worked at Auchlochan collieries, Coalburn, from 1915 until her marriage in 1920:

> Despite being such dirty wurk, we were a cheery lot. We never got ony protective clothing. Oor wurk wis pickin' the dirt frae the coal. When we had collected a lot o' waste, we loaded it into hutches that took them to the dirt bings. It wisna dangerous work we did and we seldom had serious injuries. We never had pretty nails because they were often blackened from dunts on the fingers from lumps of coal which fell on them.[23]

Barbara Marshall left school at fourteen to start work at Fordell Pit, in Fife, in 1914:

> I started work straight from school and did everything on the pit head. I can't remember how many girls worked there but there were two or three tables for different types of coal, and wagons for every different lye. They got weighed further down and and all went out on different lines. We had to set the wagons, climb up and and sort the coal so it wouldn't fall off. We started at six in the morning and worked till two. We took the redd off the coal with a wee pick and flung it aside, the rest went on the table. Some of the hutches came up the pit and they weren't weighed because they thought there was too much dirt in them. Some girls were chosen to pick the worst of the dirt out and the men didn't like it but it was the girls' job. Then the wagon went through to the tumbler and the coal came onto the tables. It was very noisy. You couldn't hear your own voice. We got one and three a day (1s 3d.). There was a pithead gaffer and a table gaffer. I did plenty of work on the pithead, put hutches on to the

Plate 46.
Overleaf. Sorting coal at pit head, Larkhall district.

cage, pulled them up the cage and put them through the tumblers. I never wore trousers in my life, we had tackety boots, long skirts and jumpers and a shawl over our heads to keep the dust out of our hair. When you finished at two o'clock you put on a clean apron to go home. You changed your pinny and washed your hair every day because the coal dust got everywhere.[24]

Margaret Davie started at Prestongrange colliery when sixteen, after spells in domestic service and farm work. At this colliery, only a few girls were employed at the pit-head alongside lads and men. Margaret did wagon work at the sorting sheds and, less than a year later, drove a pony and wagon, tipping mine waste on to a bing from a high platform.

Well, we wrocht the tumblers on the pitheid … the hutch went in and you stepped on it and turned it, and a' the coal fell doon at the other end. Ye pit the hutch away oot the road – another yin comin', ye see. An' ye'd tae keep this tumbler goin'. And the wagons underneath wis catchin' the coal … there were laddies on the tables pickin' the stanes out o' thir coal. And then after that there were yin wi' what they cried redd. And then there were a big long scaffold for the redd, and the redd wis taken up. and wis put ower the bing. Well, ah had a horse tae drive for, oh, over a year. Ah wis on this long, long scaffold masel'. Oh, ah wis seventeen past then. So ah wis on the tumblers first and then ah wis put on tae the horse. Ma wages when ah started first on the pitheid wis nineteen shillins. Mind ye, it didnae gaun very far, but it wis a big improvement on what ah'd had before. That wis still durin' the First War, of course.
    And me startin' at six o' clock in the mornin' ah got hame for ma breakfast aboot quarter tae nine. Frae ten o' clock ye worked on tae two o' clock Then well, if maybe some lassies, the like o' some o' the brickwork lassies, were wantin' tae gaun tae the dancin' – there were aye dancin' at the Drill Hall, ken, on a Monday and a Thursday or Monday and Friday – and yin wid say, 'Margaret, go and take a shift for iz,' tae get tae the dancin'. Well, they had tae pey me for their shift.[25]

The young Margaret Davie was apparently fit and high-spirited, took to pit-head work and was not averse to an additional shift in the colliery brickworks. 'But ah'll tell ye, ah enjoyed it. Ah never shirked ma job. Ah wis strong enough tae dae it. It wis jist a maitter o' hard work if ye wis anxious tae dae it.'

Barbara Marshall's parents did not want her to do pit-head work. As such work was not considered respectable and associated with disappointment and even shame to some parents, her miner father reputedly said, 'No girl of mine is going to work on the pit head', but she was determined to do so. Upon leaving school at thirteen or fourteen, few occupational opportunities existed for the daughters of miners in their own communities or elsewhere. In Fife, for instance, the main jobs were pit-head work, dressmaking, domestic service and textile-mill work. Apparently, there was a notion that these jobs had different worth and status, depending on the nature of the manual labour involved and the overall working conditions. The dressmakers placed themselves at the top, and considered themselves better than the domestics, who in turn looked down on the mill hands. The pit-head and brickwork girls were in the bottom category, their work and character bearing the stigma of coarseness and degradation.

However, Margaret Davie was probably not alone in having already experienced domestic service and rejected it for the pit head. As a servant, 'ye were at their beck and call' day and night, a slave on a pittance of a wage. 'It didnae suit me. Ah jist left there' after nine months; whereas, at least, her pit-head job was finished at 2 p.m. She had Saturday off on alternate weeks, had time to herself and had more of a wage to take home to her mother.

There was, of course, another main task that had always been readily expected from the daughters of miners who remained at home, whether or not they were in paid employment; and that was to help their mothers in the endless duties of house and home, most often in cramped, insanitary and overcrowded conditions.

Mining Communities, Standards of Living and Collective Action

For four months in 1901, a sympathetic young American, Kellogg Durland, stayed and worked as a miner in Kelty. The short book relating his experience and observations is an interesting and useful record.[26] He lodged with a mining family in a recently built two-room house owned by the employer, the Fife Coal Company. One of a brick row, this kind of house is recognisable as a Scottish 'room and kitchen' dwelling, only that it also had a triangular roof with an attic used for living space. There were no less than fourteen people living in this company house at the time of his visit. The family of nine slept in the 'room' (parlour) and in the 'kitchen' (the living room area). He and another four boarders were accommodated in the attic. The beds in the attic were rotated to accommodate the different shift pattern of the men. The service for the boarders included food, washing and mending of clothes.

Durland could well ask why large families 'should be obliged to cramp into two small rooms' when their members were employed 'by a Company issuing such enormous dividends' to its shareholders. 'Add to this the evils, perhaps at the present time unavoidable, of the lodger system whereby people occupying two or three rooms take in several lodgers, and the answer is obvious. Overcrowding is inevitable under the present system in towns and villages that are growing as rapidly as Kelty and certain other Fife villages.'[27]

Like the mushroom-growth settlements of the Monklands in the 1830s and 1840s, Kelty was one of several boom towns created by the expansion of mining in Fife at the turn of the century. The pattern of development and the accompanying social pressures are a familiar story. It was in the middle of a new mining frontier whose work opportunities attracted many incomers from other parts of east and central Scotland. However, although the large coal companies did build houses, most often they failed to create or invest in sufficient housing accommodation for their enlarged workforce. Moreover, in those circumstances of housing shortage, companies still expected mining families in their tied houses to take in lodgers and boarders. Durland confirmed that this was the case in Kelty, where the Fife Coal Company owned around two-thirds of the housing stock. In this typical single-industry village of 5,000 people in 1900, he noted the difference in living standards, morale and outlook among the miners and small tradesmen who were owner-occupiers of the other houses. These houses, separate from the rows, were taken on the basis of mortgage payments to a building society, and were occupied mainly by mineworkers whose wages and prospects were better than those of their fellow workers. In contrast to the overcrowded conditions and dreary surroundings of the many two-room company houses, he found this minority of other houses 'for the most part singularly attractive from the outside, and cosy and homely within'. The occupiers took 'pride in their home, and tending the garden becomes a supreme pleasure. The miner's wife seizes the opportunity to give rein to her womanly instincts which appear when she is left free to make the home in her way. In short, it raises the whole standard of life.'[28]

Such scope for 'a little taste and individuality' was always impossible in a grossly overcrowded company house in a brick row with filthy outhouses and desolate back yards. Here, home life was repressive for everyone concerned, but especially so for the women and girls who were also subjected to male domination. This dimension of female subordination was, of course, neither new nor exceptional to mining communities; but it is emphasised, although explained with some difficulty, by Durland in some revealing passages.

> I was at first surprised, but gradually came to take as a matter of course the servitude of women. Their slavery to the men was almost universal throughout the district. The men were looked upon as the wage earners and the lives of the women were given up to making them comfortable. Not once can I remember of the women eating their meals with the men in our home. In some houses where the families were smaller and the tables larger it might have been possible for the women to eat with the rest, but in our house to have made room for them would have meant crowding and cramping the men. Any suggestion of inconveniencing the workers would not have been tolerated at all. There were two big easy chairs in the kitchen (which was our common room) and if any of them chanced to be occupied by one of the girls or women when the men arrived it was instantly left for one of the men to drop into. It was a common thing for the men to demand that their pipes be filled by one of the women. I have seen a son of one or two and twenty order his mother across the room to get his pipe which was on a shelf directly above his head a few inches out of his reach from the chair where he was sitting. All the time the men were at home the women would hover about ready to be instantly commanded for the most menial services.

Aware that his remarks were controversial in this regard, he chose his words carefully after checking over the accuracy of his observations with some of the miners. However, one of them, who had 'a lifelong familiarity' with such matters, was indignant at Durland's use of the word 'slavery' to define the women's behaviour towards them, insisting instead that a more appropriate word was 'devotion'.[29] Durland then admitted that he had not understood the complexity of gender relationships within mining households, where the women made so many sacrifices to care for the male breadwinners who were in constant danger of their lives. Fundamentally, all of them under the same roof were oppressed by a system in which the mining companies grossly exploited the wage earners and their families. These circumstances, however, could not hide the double oppression that the women faced in supporting and sustaining the male workers at home.

This view of the vital role of the homemakers in the most difficult living conditions is endorsed in many contemporary accounts, as in this tribute by an Ayrshire miner who was born in 1892 and raised in Dreghorn in a small company house without an indoor water supply. Bringing up five of a family, with her husband and three sons down the pit, his mother

> was just a slave, like the rest of us. When we came home from the pit, my mother and young sister had a large wooden tub placed on the centre of the kitchen on the stone floor, with hot water to wash ourselves. The water had to be heated on the fire in the kitchen in large pots, which were hung on swivels over the fire. This meant that we had to sleep, wash, eat, and live in the kitchen. If we wanted a bath, we had to go outside and round the back of the houses to the wash house. Our wives or mothers had to put the boiler fire on and fill the boiler with water.
>
> My mother had to wash in the wash house. She lifted the heavy clothes out of the boiler with a long piece of stick or clothes pole into the wash tub, then with her bare feet, holding up her skirts, would tramp on them, then wring them and hang them up to dry.
>
> My oldest sister did all she could when at home, but she had to go into country service. The hard, slaving work was left to my mother and my younger sister, and she became a woman before her time owing to all the hard work. Mother would sit up most of the night to darn socks and mend pit clothes and dry them for the following morning. My sister would help by scraping the mud off the pit boots, for we used to come home from the pit with our clothes and boots all clay and mud.[30]

This typical experience of harsh domestic conditions in early twentieth-century Lanarkshire is vividly recalled by Susannah Pate, whose father and brother were miners at Coalburn. The family lived at Thornton's Rows, Kirkmuirhill:

> I was born there and these were two raws of six hooses. We had a room and kitchen wi' 'holes in the wa' or 'set-in beds'. We never had water closets. It was dry closets roon the back o'the raws and they were shared wi' ither faimilies, all wi' a lot o' weans. A man used to come roon wi' a horse and cairt to empty the buckets in the close and the middens where we threw oor rubbish and ashes. In summer, folk would put up fly catchers in the dry lavatories and they werena right up before they were black wi' flies.
>
> Her weekly wash was a marathon for ma mither. It was two shared washhooses for the twelve faimilies. Mither stairtit the nicht

before. The boiler and a big tub had to be filled. Usually the water was brocht frae the pump but mither had a barrel at the side o' the entrance door and a rone pipe ran rainwater frae the roof into the barrel. It was 'soft' water so was suitable for washing clothes.

Under a bed was stored a big bath that ma faither and brither used when they came hame dirty and before they had access to pit baths. Before they walked in, mither would be boiling kettles o' water. They were poured intae the bath placed in front o' the kitchen fire. It was very elementary wi' very little privacy.

We had nae runnin' water in the hoose, so all the water had to be cairrit in and cairrit oot. The wurkin' claes o' miners were often sodden, sometimes even frozen stiff, when they came hame frae the pit because o' wurkin' in wet conditions. These had to be dried for

Plate 47.
Houses and muddy exterior at Allanton pit village, near Hamilton.

Plate 48.
Overleaf. Lumphinnans back-to-back rows on a clear day.
(*National Coal Board, Scottish Area*)

puttin' on the next mornin'. The moleskin troosers had to be scraped o' dirt and so had the heavy boots. These were the jobs that maistly fell to the mither and dochters in the hoose.[31]

Abe Moffat, born in 1896, and raised in Lumphinnans, Fife, recalled the housing conditions:

> My native village was typical of the mining villages at that time, as most of the houses had only one or two rooms. Some had brick floors and no gas or electricity; only paraffin oil was used for lighting … and the smell of oil was always there.
>
> Many mining villages depended on water from the pits. No wonder at that time that miners' families paid almost as much for lemonade and soda as they did for rents. The typical drainage system was an open surface channel or gutter which ran quite near to the houses. The stench from the dry closets, or privy middens, as we called them, was deplorable, especially in the summer time, when it was a month or more before the refuse was removed. These were the places where children had to play, and it is not surprising that disease was common in mining villages as a result of such inhuman living conditions. Flies were prevalent and, dipped in filth from the privy middens, they would invade the houses and settle on food, milk, jam and bottles.
>
> All this showed the deep disregard on the part of the employers and local authorities for the needs of human beings, and it is no exaggeration to say that miners and their families were looked upon as sub-human beings whose only purpose was to produce coal and profits for the coalowners. Yet miners' wives were always concerned about the cleanliness of the house. How they were able to keep houses so clean inside under such conditions is beyond anyone's imagination.[32]

The individual snapshot impressions of living conditions and domestic roles can now be compared to the known wider profile of housing conditions within mining communities throughout Scotland. It is the conclusive verdict of social historians of the mining community that colliery-owned housing in Scotland into the twentieth century was generally inferior to tied housing stock in other regions of the British coalfield. The horrifying housing conditions in most Scottish mining communities were exposed in graphic detail in several important enquiries, including county medical officer surveys conducted in 1910, in the evidence and report of the Liberal Government's Royal Commission on Housing (Scotland) between 1913 and 1918 and in the Sankey Commission of 1919, which examined the earnings and housing conditions of miners in the industry's post-war crisis. The main findings of these enquiries, particularly of the Royal Commission, are summarised below. They reinforce the evidence of the personal recollections in this chapter.

The typical small, single-storey houses in the miners' rows and squares which had been built before and after the mid nineteenth century in company villages and settlements in rural areas had long since degenerated into squalid slums and disease-infested death traps. Over the years, some of these abominable houses had been abandoned as nearby pits had closed, and had been condemned as being unfit for habitation. Yet, most were still inhabited by mining families, sometimes against their will, as a condition of employment.

By the early twentieth century, over a third of mineworkers in Scotland were living in company houses, mostly in rural and landward areas, while the majority rented private accommodation in the towns, close by the collieries or within travelling distance. A further breakdown of the figures for company-owned housing in early twentieth-century Scotland reveals that up to 70 per cent of miners in Ayrshire and the Lothian counties inhabited this

Plate 49.
Smeaton pit village, Midlothian, with two-storey houses and outside stair to upper floor. (*Collection of the Scottish Mining Museum Trust*)

category of house, which was mostly room and kitchen. The corresponding estimate for Fife miners fluctuated around 50 per cent. In the west-central region of the coalfield, comprising Lanarkshire, Stirlingshire and Dumbarton, the proportion of miners in company housing was lower, at around 30 and 40 per cent. The North Lanarkshire towns were well populated with miners, and in the two smaller counties miners had obtained houses in the private rented sector in Stirling, Falkirk, Larbert, Denny and Kilsyth. In all these localities there was considerable travel-to-work activity among the miners, on foot, and by bicycle, bus, rail or tram.

There were significant local and regional differences in the housing conditions of miners throughout the Scottish coalfield. In general terms, the worst housing and environmental conditions were in Ayrshire and Lanarkshire, which contained the bulk of the older, most dilapidated and insanitary housing stock. The best of a very varied housing provision was found in Midlothian and Fife, where investment in colliery expansion and future prospects was at times accompanied by improved new houses to accommodate thousands of incoming workers. Notable examples of the few quality company houses were to be found in the 'model villages' of Newtongrange, High Valleyfield and Dunbeath, Buckhaven. While mining families in new houses in Midlothian were enjoying the benefits of two or three apartments, a scullery and an indoor water supply, even larger housing was common in Newtongrange, which, before 1914, was reckoned to

Plate 50.

Better-quality housing for miners, Newtongrange, early twentieth century. (*Collection of the Scottish Mining Museum Trust*)

contain 'probably the best miners' housing in Scotland'.[33] In this strictly controlled company town of neat cottages, rows and terraces and well-kept gardens, a significant proportion of the houses had four or five rooms, a scullery and kitchen and indoor water closet, although bathrooms were not yet provided. The 'carefully designed new village' at High Valleyfield, near Culross, included houses arranged in crescents, with plenty of open space and gardens attached. The houses were typically two and three apartment, with scullery and flush lavatory, and some had baths with hot water from the kitchen range back-boiler.[34] Although less commodious, new houses at Buckhaven departed from the monotonous rows that formed the principal pattern of colliery housing in Fife. They were two-storey dwellings, with external stairs, the tenant on each level being supplied with lobby, living room, two bedrooms, scullery and water closet.[35]

Across the Scottish coalfields, all but a few mining families were either too poor or unsettled to contemplate house ownership, far less to realise it. However, as in Kelty, and in east Fife around Wemyss, some of the more prosperous and settled mineworkers owned their houses, or were prospective owner-occupiers. An estimated, low 5 per cent of mineworkers in West Lothian owned their houses, while at the opposite end of the scale, in exceptional circumstances discussed earlier, 155 out of 201 Leadhills miners lived in their own houses in 1909.[36] Another conspicuous area of miner owner-occupancy – around a third of miners – was in and around Larkhall

Plate 51.

Postcard of Leadhills village, showing houses of miners and smelters. (*Wanlockhead Museum Trust*)

and Stonehouse, where the tradition of independent collier respectability and investment in building societies was long established in an area otherwise dominated by rented colliery housing. Compared to mining families who were tenants of employer landlords or other property owners, and who never knew when they would next have to flit, settled miner owner-occupiers were much more likely to take greater care and responsibility for the appearance and upkeep of their properties, including sanitary standards. However, this is not to deny the heroic efforts of countless women living in insanitary company rows and other deficient rented properties who struggled against all the odds – amid poverty, many pregnancies and the distress of frequent infant illness and death – to maintain domestic hygiene and decency, and their own sanity.

The housing enquiries revealed that extreme overcrowding was rife in mining and industrial settlements, large families of tenants being packed into one- and two-room dwellings. By 1920, local authority enforcement of building by-laws and public health regulations in mining and industrial towns and burghs had begun to tackle the shocking sanitary condition of the older houses. By 1914, in congested burghs like Motherwell and Wishaw, most of the filthy, stinking and leaking privy middens and open ash rubbish pits only a few feet away from the houses had been removed. More regular scavenging, installation of water closets and piped water had begun to encourage greater hygiene and to lower the awful death rates, which were nearly one in seven among infants in such urban areas. Thereafter, public-health standards showed some improvement, although the major blights of overcrowding and serious shortage of housing remained.[37]

Living conditions, for the most part, were even worse in the mining villages and isolated rows within county council areas, where provision and enforcement of regulation standards of housing and sanitation were hardly evident. Neither coal companies nor other principal ratepayers could be persuaded to defray the costs of a proper water supply, drainage and sewage to scattered and remote mining villages. The problem of poor sanitation was particularly acute in Ayrshire, as depicted in damning evidence submitted in 1913 to the Royal Commission on Housing by the leaders of the Ayrshire Miners' Union. In the county with most scattered mining settlements in Scotland, they concluded that 'the Public Health Acts, so far as the Ayrshire mining rows are concerned, are practically dead letters'.[38] Failure to implement the law was blamed on the propertied interests that dominated the Ayrshire County Council and on the reluctance of sanitary inspectors to offend these men, who were their employers.

By 1900, the Scottish miners, represented by their trade union organisations and in political campaigns, had begun to feature prominently in the long struggle for reform and overhaul of housing and health provision for their own families and for the wider working class. According to this political stance, the state and local authority had to take responsibility for housing the workers, backed up by financial assistance, and thus bring to an end the miserable housing and health record of companies and private enterprise who obstructed tolerable living standards in mining communities. At a personal level, the campaigns on these linked issues were as much about saving and improving the lives of women and children as about the welfare of the miner coming off shift and in his home environment.

The former miner, housing crusader and Clydeside socialist, John Wheatley, who was to become Health Minister (also with responsibility for housing) in the first Labour Government in 1924, had grown up in bleak poverty with parents, seven brothers and sisters and sometimes two lodgers, in a single-room miner's cottage at Bargeddie, in Lanarkshire. Domestic water was carried from a pump a hundred yards away, and the family had to walk fifty yards to reach the dry privy. Fired by indignation at such conditions, Wheatley resolved to find solutions to the social miseries that befell so many mining families. In a campaigning pamphlet of 1909, he wrote: 'There is no section of people who require a bathroom more, and there is surely none who have less facilities for washing. The lot of the women is, in many respects, harder than that of their husbands.'[39] Without a bathroom and fire-generated hot water, public wash houses and workers' canteens, pit-head baths with changing and drying rooms and adequate living space in the house, he insisted that there could be no end to the back-breaking everyday drudgery of the women who had to stoke coal fires in wash-house boilers and house grates; carry, fill, heat and empty the contents of pails and tubs; scrape, wash, wring and dry pit clothes; make meals for shift workers and large families; and somehow also keep a tidy, clean house.

The Coal Mines Act of 1911 had legislated for pit-head baths but Abe Moffat explained,

> As we found from experience, legislation always provided loopholes to allow the employers to avoid their responsibility. In the Act there had to be a majority of two thirds by ballot vote of the miners before an employer was compelled to erect baths providing the miners paid half the cost, and the owner was not bound to provide such accommodation if the estimated cost of maintenance exceeded 3d. per week for each workman. Another part of the Act exempted the employers from erecting baths if the mine employed less than 100 workmen. So

these escape clauses allowed the coalowners to escape the responsibility of erecting baths.[40]

The first pit-head baths in the country were opened in 1916, at Douglas Castle Colliery, where the Coltness Iron Company provided them directly as a welfare measure.[41] However, there are no other known examples of this provision before the colliery welfare schemes of the 1920s and 1930s, and so the widespread demand continued for both pit-head baths and adequate modern houses with bathrooms.

The appalling conditions of health and housing had to be addressed at all levels of the political process. Increasingly, from the turn of the century, there was a growing realisation that the desired changes would not be won until the stranglehold of reactionary propertied interests had been broken in local councils and in Parliament. Although weak during the 1880s, trade unionism among miners already existed to protect and advance their interests in the workplace. However, to build a powerful force to campaign effectively for rights at work and improve living conditions within mining communities required strong trade union organisation and funds. Such resources would be used not only to support industrial action and stand up to the employers, but also to contest elections and win political representation for the world of labour. How mineworkers in Scotland faced those challenges is the final theme of this chapter.[42]

Industrial Action and Politics

Before 1900, the growth of trade union membership and stable organisation among miners was slow and precarious. Permanent county unions existed only in Fife and Clackmannan and Mid and East Lothian, although their influence was slight in many company villages. Yet, in these counties, the employers recognised the union and conciliation boards kept disputes to a minimum. In central and west Scotland, where industrial relations were more primitive and turbulent, only a small fraction of miners were union members, although non-unionised miners were often capable of successful workplace action over hours, wages and conditions. Nevertheless, internal divisions among miners, including the demoralisation resulting from low wages, defeats and setbacks, and the implacable opposition of employers, managers and contractors towards unions and wage bargaining, impeded organisation and unity. In most areas, union activists were liable to be fired and blacklisted by employers and many miners were afraid to belong to a union or be seen to cause trouble for fear of losing job and house.

Having to contend with highly authoritarian, coercive owners and
managers was nothing new for miners and their unions. The business profes-
sionalism and undoubted mining expertise of the better-organised modern
companies was often accompanied by ruthless discipline in the workplace
and in the company villages they owned and controlled. From the mid-1890s,
the authoritarian regime operated by the Lothian Coal Company at its
Newbattle collieries among the workers and families at Newtongrange and,
to a lesser extent, in its surrounding settlements, has passed into legend as
one of the most notorious examples of its kind. Mungo Mackay was, by all
accounts, a tyrant of a manager who, for almost fifty years until his death in
1939, exercised a virtual reign of terror over every aspect of life and work
within his domain. He did so in the interests of efficiency of production and

Plate 52.

Mungo Mackay in
masonic regalia.

profit for his masters and shareholders, and pursued a strict paternalist policy in ordering the affairs of the mining community, including the building and allocation of decent housing and other amenities. A formidable personality and an expert mining engineer, a staunch Protestant and Freemason, he ruled by intimidation and fear, relying upon managers, oversmen and contractors, and a spy network of workers and villagers, to keep him regularly informed. He exercised direct powers of hire and fire, and used his control over the high percentage of workers who occupied colliery housing, especially in Newtongrange and Rosewell, as a constant threat to worker tenants who persisted in disobeying his will. All miscreants were summoned to the main colliery office at Newtongrange, where he acted as judge, jury and executioner. Punishments included fines for misdemeanours such as failing to keep a tidy garden and house, being drunk and disorderly, or otherwise creating trouble with neighbours. More serious offences above or below ground resulted in dismissal. Even the consumption of alcohol was strictly monitored in the large, and only, public house in the village, and strict discipline and fines were imposed here for bad behaviour.[43] The Dean Tavern, built in 1899, was a Gothenburg public house (or 'Goth' for short), owned by the company. The Gothenburg scheme of public houses had originated in Sweden in the 1860s, allowing registered bodies, including town councils, to run licensed premises on reformed lines, with the aim of curbing excessive drinking and providing social amenities and community benefits from the profits of the sale of alcohol and food. There is no doubt that this scheme had its attractions for coal companies, as they could control the drinking habits of many of their workers, and also save themselves the expense of providing facilities in the mining communities. In effect, the miners who used these public houses paid for the amenities out of their own pockets and wages. Thus, in Newtongrange alone, public house revenues over the years paid for a bowling green and pavilion, a park, a cricket pitch, a cinema (1915), grants to clubs and societies and at least one community centre. A board of management ran the scheme, with miner representation, but dominated by company officials, and headed by Mackay himself from 1917.

Many Goths were established in mining villages and in coal towns, including over twenty in Fife before 1914 (Cardenden had three, and Kelty four); several in the Lothians, and one in Stirlingshire and Ayrshire. All maintained strict codes of behaviour, including a ban on betting and gambling. The village Goths were almost always company dominated, and at Fallin, Stirlingshire, the pit contractors were majority shareholders in the local Goth, often paying out wages on the premises to their miners from

Polmaise and neighbouring collieries.[44] In company villages like
Newtongrange, before the inter-war years, the official trade union presence
was generally weak, and prolonged local industrial disputes were almost
unknown, although all such mining communities were drawn into the
national-strike confrontations from 1894 onwards.

During this period, among mining communities in the west, it is difficult
to detect the existence of authoritarian company regimes as extreme as that
at Newtongrange. However, at Bothwellhaugh, the mining village built and
controlled by the Bent Coal Company for their new Hamilton Palace
Colliery, similar features operated. From its beginnings in the 1880s, it was
fiercely anti-union; it managed and ran a store and permitted no other
traders; and it had a monopoly of company houses, which accommodated
almost the entire 1,200 workforce and their families before 1914. The
colliery manager and chief cashier, Stewart Thomson, allocated the housing,
and was alleged to have reserved the best-quality houses for compliant
workers who were not trade unionists. Company policy prohibited the sale of
liquor in the village, and was tough on drunken behaviour. Thomson's spies
and stooges also kept him fully informed of everyday life in the 'Pailis' (as
Bothwellhaugh was known locally, in ironic contrast to the grandeur of its
aristocratic neighbour, Hamilton Palace, immediately across the Clyde).[45]

The Leadhills Company also adopted a fierce anti-trade-union stance,
refusing to negotiate on wages. A dispute, lasting for over three years
between 1898 and 1901, had prompted the lead miners to join the
Lanarkshire Union of Mine Workers. They received strike money and coal,
but company intransigence and division among the miners led to men leaving
the village or being dismissed. The same painful process happened in 1909,
when the company staged a lockout. In an astonishing display of labour-
movement support, a solidarity demonstration of around 5,000 people rallied
to the remote village on 9 June, although the company again refused to
budge on trade union recognition and negotiation. Labour relations
continued to deteriorate for years, and the end of lead-mining and smelting
was in sight as ore finds diminished along with the workforce.[46]

In the west, only around 10 per cent of mineworkers in Ayrshire and
Lanarkshire were union members in the 1890s. Even in areas such as Blantyre,
with its big mines, its visibly young, large and unruly workforce and its
growing reputation for militancy, the number of dues-paying members was
abysmally poor. In the western coalfield generally, miners were frequently on
the move, which acted against stable union membership. Also, before 1914,
union recruitment was still largely confined to hewers and underground face
workers who were on tonnage and piece-rates, instead of developing into a

general union to embrace other mineworkers, including ancillary and surface workers of both sexes. In some areas, there was also an intense parochialism, tinged with sectarian feeling, which had a divisive effect on worker solidarity. For example, the staunchly Protestant and Orange mining district of Harthill and Benhar did not welcome Catholic workers or incomers, and for years was averse to seeing its trade union funds go outside its home area to support disputes elsewhere.[47] Moreover, the Scottish miners had to suffer gruelling industrial struggles before their position improved by 1914.

The earliest major confrontation in this period was in 1887, when parts of the coalfield and the shale areas were convulsed in a bitter reaction to wage cuts. The shale miners went on strike and then were locked out by the employers. Industrial action was centred in the Broxburn area, where miners stayed out the longest, for twenty-one weeks. The Broxburn Oil Company ordered wholesale evictions from its houses, but families rallied round to give temporary accommodation. Local women were also involved in harassment of blackleg labour; violent scuffles occurred; police were frequently in action; and soldiers arrived from Edinburgh to reinforce the police.

Still, the employers conceded, and the shale miners' union was recognised, a nine-hour day agreed, and previous tonnage rates restored, which amounted to a significant victory for unionised labour in the shale industry. The notable refusal of unemployed coal miners to be strike-breakers in the shale-mining dispute undoubtedly helped towards a successful end on this occasion.[48]

In the west, Blantyre was the main storm centre of the 1887 episode, when coal miners were implicated in communal rioting as well as violent picketing. In this growing frontier area, relations between police and community were already badly strained. Law and order forces were fully mobilised, including police reinforcements from Glasgow, and troops on horseback. Order was restored by police baton, with many arrests.[49]

In the first national strike, against savage wage reductions in 1894, around 70,000 mineworkers were involved, although only 30,000 were trade union members. After fourteen weeks, the struggle ended in defeat after funds ran out; resolve weakened as the influence of contractors and sectarian bitterness took over, and hardship forced a return to work. This episode of struggle was also marked by a proliferation of communal soup kitchens, mass picketing, female harassment of blacklegs, escort of blackleg labour by police and several confrontations, particularly in mid Clyde valley communities, between pickets and police. Soldiers were sometimes called out in reserve, as at Motherwell on 18 September. Union membership fell away as demoralisation followed defeat.[50]

Yet, despite such reverses and internal divisions within communities, mining trade unionism and its influence underwent a phenomenal growth after 1900. With only a few thousand members in 1890, the Scottish Miners' Federation, comprising eight county unions and including shale miners, had 110,000 members by 1914. The main reasons for this growth are briefly as follows. In 1900, the main employers in the west agreed to conciliation boards and to sliding scales of wages according to the state of the industry, which at least allowed for union recognition and gritty negotiation by both sides at county level. In 1894, an important piece of labour law was enacted, whereby managers had to recognise the right of miners to elect and employ checkweighers at the pit head. This hard-won victory contributed to the effectiveness of collective action at pit level and helped build confidence in the union. County unions increased their members and funds. They employed agents and officials to fight compensation cases in the courts, to negotiate with employers and otherwise represent their members when wages were failing to keep up with inflation before and during the First World War. The Trade Union Act of 1906, enacted by the incoming reforming Liberal Government, restored the right to peaceful picketing and removed the threat of confiscation of union funds for employer damages arising from disputes, thus improving the status and bargaining position of miners' organisations. Organised miners also took a more determined stance against non-unionism, although resistance to contractors and contracting was less successful, particularly in the mechanised pits.

In March and April 1912, the Scottish Miners' Federation was better placed to stage the first national strike over a minimum wage for all grades of mineworkers. Again, this was a bitterly fought episode, characterised by increased levels of police retaliation and of militant, direct action to prevent strike-breaking. Most of the major incidents were in Lanarkshire and other west-central areas, but Fife miners, supported by their women, were involved significantly in militant action for the first time. During this strike, it was clearly evident that Lithuanian miners had established their trade union credentials, notably as successful pickets in Lanarkshire and West Lothian. Another notable, new, social feature of the long strike was political pressure exerted on school boards to grant free meals to the children of strikers. For instance, in Motherwell, 1,800 children received two meals seven days a week, setting an example that would be repeated during the 1921 and 1926 lockouts.[51] The Government conceded the minimum-wage principle and ordered its implementation at local level, although the outcome in terms of raising real wages proved disappointing and continued to cause resentment into the war years and beyond.

Plate 53.
Overleaf. Scavenging for coal during the 1912 strike. (*East Dumbarton Council*)

During the 1912 dispute, Robert Smillie, the Scottish miners' leader, addressing a mass meeting in Motherwell, had urged miners to continue the campaign to secure a decent statutory minimum wage not only by using their industrial muscle, but by political action through the ballot box in support of Labour representation in local and parliamentary elections. However, since the days when Keir Hardie had first stood unsuccessfully for Labour in 1888 at the Mid Lanark by-election, Smillie's keen advocacy of harnessing the voting and trade union power of the miners to the electoral fortunes of Independent Labour politics had always proved an uphill task. The millstones of sectarian religion and ethnic divisions among the workforce – in particular, the Catholic Irish allegiance to the Liberal party and the politics of Home Rule, and the militant Protestant and Orange allegiance to Unionist politics and the Conservative party – had hindered a Labour breakthrough in mining constituencies in west-central Scotland. The Liberal Party was the dominant political power in the land from the middle of the nineteenth century until 1918, and was supported by many of the coal-owners. Inasmuch as miners had explicit political views, most were attached to some form of liberalism or reformism. They tended to support the Liberal Party in elections, rather than declaring for labourism or socialism, although these influences were becoming more apparent among mineworkers. The formal affiliation of most of the county unions to Labour before 1914 signalled a difference in the electoral behaviour and political attachment of miners and their organisations.

Plate 54.

Left. Robert Smillie.

Plate 55.

Right. Keir Hardie, 1892.

Miners' leaders like Smillie had started out as checkweighmen, before becoming union officials and agents. They had then cut their political teeth as candidates for burgh, parish and county councils, before seeking election to Parliament on behalf of the emerging Labour Party or more over socialist organisations. They did so in constituencies that had a large number of male miners who were entitled to vote, as a result of the extension of political rights in 1884 and 1885, and where the funds of the union were most vital in fighting elections. However, these efforts at winning elections had met with modest success at local level but, with only one exception, all challenges at parliamentary constituency level had failed before the 1918 General Election. This sole victory for Labour was in West Fife constituency, where Willie Adamson, the cautious, moderate, regional miners' leader, was returned in 1910 and held the seat for many years. It was perhaps no coincidence that this was the only constituency in Scotland in which miners formed the majority of voters. All mining constituencies were contested in 1918 for Labour, with great expectations. Although the voting counts were often high, victories were achieved only in Ayrshire South, where veteran miners' leader James Brown was elected, and at Hamilton, where local miners' agent Duncan Graham was also elected. The Labour landslide in mining constituencies was to be postponed until the General Elections of 1922 and 1924.

Within mining communities, notably in Lanarkshire and Fife, strong undercurrents of discontent had emerged between 1917 and 1920. Anti-war feeling increased as prices leapt ahead of wages and mine-owners were seen to profit from a war that was killing and maiming far too many volunteer and conscript miners in the slaughterhouse of the Western Front. A minority grouping of young left-wing militants was also coming to the fore to challenge the policies and stances of the older and more cautious miners' leaders. This agitation was led by a growing Miners' Reform Movement with demands for workers' control of the mines, a guaranteed living wage, a six-hour working day, and a five-day week. These proposals were promoted at mass meetings before and during wages strikes in 1919 and 1920, and in political education classes led by left-wing teachers and activists within mining communities. They were widely influential among rank and file miners, especially in centres of the movement within Lanarkshire such as the Hamilton-Blantyre district, Coalburn and Shettleston; and in Fife, at Bowhill, and around Cowdenbeath and Lumphinnans.[52] Such demands for radical change of the mining industry, and for improved working and living conditions, would all be put to the test in the most bitter struggles during the prolonged crises of the inter-war years and beyond.

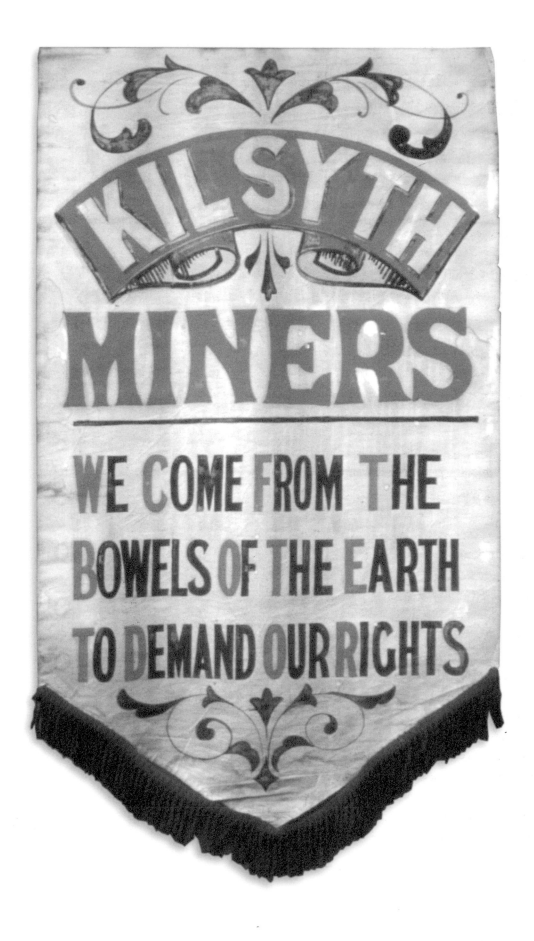

Toil and Trouble in a Declining Industry, 1920–1946

The Mining Sector

During the First World War, amidst rapid rises in the cost of living, miners had won wage increases, due to the great demand for coal and reluctant government recognition of the industrial strength of trade unionism in the coalfields. As an organised body, coal miners emerged from the war more powerful than ever before, and were in the vanguard of the working-class movement. In 1919, new demands – for a living wage, a six-hour day, a five-day week and nationalisation of the mines – expressed an optimistic determination that further improvements in working conditions could be wrested from Government and owners. To stave off further unrest among the miners, the Tory–Liberal coalition government conceded a seven-hour day. The Sankey Commission, appointed to investigate the industry, recommended nationalisation and awarded a wage rise, which was worth up to an extra £10 over a four-month period. A Shotts miner recalled that some young single miners felt flush with the 'Sankey money', got married and set up house in rented rooms.[1]

However, conditions and prospects changed rapidly for the worse when the brief post-war boom collapsed in 1920. The coal industry was plunged into crisis as the bottom fell out of the export market, and tumbling prices hit production in the Fife and Lothian pits, which largely depended on this market. The west-central coalfield was even more deeply affected by the sudden downturn in orders for shipbuilding at the huge Clydeside yards, resulting in much reduced demand for coal and iron ore to fuel the furnaces of the iron and steel works that made the ship hulls and boilers. This crisis devastated the coal-mining and heavy-industry communities of North Lanarkshire, especially the Motherwell–Wishaw heartland, where the steel plants and their large workforce were made redundant. There was a partial recovery in 1924, but the depression in heavy industry lasted for over ten long years. Renewed conflict in the mining industry in 1921 had shut off vital coal supplies, reinforcing the already large-scale local unemployment among mineworkers.[2]

Plate 56.

Kilsyth miners' banner. (*North Lanarkshire Council*)

Mass unemployment in the areas of coal mining and heavy industry during the inter-war period is usually associated with the peak years of the economic slump in 1932–3. However, the coal and steel communities of North Lanarkshire were especially hard hit in the crisis of the early 1920s, when unemployment levels almost matched the record figure of 50 per cent for the early 1930s. The position was even worse in all villages where coal mining was the sole occupation, as most men and boys were out of regular waged work for long periods, and even for several years during the 1920s and 1930s.

In March 1921, the Government rejected the recommendation of the Sankey Commission to retain state control of the mines and handed back the pits to the private owners. This action was a shattering blow to miners' expectations of a nationalised, planned industry with enhanced working conditions. Encouraged by the government decision, the coal-owners announced sweeping wage cuts and an end to nationally fixed agreements on minimum pay and conditions. The miners refused to accept the new terms and on the very day of state decontrol of the industry, 31 March 1921, contracts of employment were terminated and the miners were locked out of work. A bitter confrontation lasted for three months before the miners were forced back on the owners' terms. This defeat was a decisive turning point, coinciding with deepening economic depression, short-time working and unemployment in many collieries. In the mid 1920s, the miners attempted to defend wages and conditions against the continuing offensive of the owners, whose labour costs were supported by temporary government subsidy in 1925. The end of this subsidy in April 1926, followed by refusal to accept wage cuts, provoked the historic nine-day General Strike of workers throughout Britain in support of the miners (3–12 May), and eventual defeat of the miners in the long lockout between May and November. The spirit and endurance of miners and of their communities in Scotland during the major episodes of struggle in 1921 and 1926 are illustrated in the next section of this chapter. Miners also took part and sometimes provided leadership in several 'Hunger Marches' during the 1920s and 1930s, in protest against mass unemployment and the degrading Means Test, which was introduced in 1931. While the national marches, involving Scottish contingents walking to London from depressed central Scotland, are well documented, regional marches were also important events, as in the cases of the Fife-to-Edinburgh and the across-Fife marches of the early 1930s, in which unemployed miners formed the majority of protesters. Those experiences were to show that many miners were not willing to be mere passive victims of powerful forces that threatened destruction of their livelihoods and, in circumstances of extreme deprivation, their very existence.

After 1926 any short-term prospects of public ownership of mines had disappeared, and the miners had instead to confront the harsh realities of rationalisation and restructuring within a declining industry, bringing massive changes to the workplace and to mining communities. Having already driven wages down to rock bottom, and able to inflict longer working hours, from the late 1920s into the depressed 1930s, the larger companies sought to raise productivity and make profits by pushing through further fundamental changes in mining operations. The principal measures included concentrating production in the more efficient pits and closing uneconomic pits, and speeding up and intensifying the work process at the coalface by increased use of machine-cutters and conveyors. Many of the older pits where shafts and haulage roads were too narrow to permit worthwhile investment in modernisation were closed altogether, joining those pits that had already been dismantled after flooding during the prolonged stoppages of 1921 and 1926. Others were closed in the 1930s, notably in North Lanarkshire, where huge tracts of waterlogged collieries between Glasgow and Coatbridge, and in the Wishaw-Cleland district, were abandoned by owners who were unable or unwilling to plan for, and share, the costs of pumping and maintenance operations. In the West of Scotland, over ninety of the less productive pits were closed between 1924 and 1937 with devastating social consequences,

Plate 57.

Boy, pit pony and lamp.
(*Collection of the Scottish Mining Museum Trust*)

and even in Fife by 1931 only twenty-seven pits were working, compared to sixty-four in 1924.

The combination of depressed industrial demand for coal and restructuring of production and work methods resulted in the loss of thousands of jobs in mining and greater exploitation of most of those workers who remained at the coalface. In the Scottish coalfields, within ten years, the number of mineworkers had fallen from a peak of nearly 150,000 to a low of 80,000 at the worst point of the slump in 1932–3. Recovery in the late 1930s saw the figure rise to 90,000, before dropping again to the 1933 level during the years of the Second World War. So deep was the economic and social crisis in the Lanarkshire coalfield that half of its workforce in mining (around 30,000 workers) was made redundant between 1920 and 1938, while Fife lost over 25 per cent and the Lothians around 20 per cent of their mineworkers.

In Scotland, overall production of coal fell from a record 42 million tons in 1913 to a yearly average of around 30 million tons during the inter-war years, although this global figure disguises the low levels of production in strife-torn 1921 and 1926 and in the slump of 1931–3. However, the rise in labour productivity accounts for the surprisingly large annual average in coal production, especially after 1926, brought about by the intensified and widespread implementation of mechanised coal cutting and haulage and by a ruthless reduction and redirection of the workforce. Ownership of the mines had become even more concentrated in the hands of a few large firms whose investment in mechanisation, more effective use of labour, authoritarian anti-

Plate 58.

Machine coal-cutting.

(*Collection of the Scottish Mining Museum Trust*)

union policies and reductions in wage rates had driven down the costs of production. Although company profits were badly hit during the early 1930s, and many small mining firms went bankrupt or into debt, almost all large companies operated at a profit by 1939, and dividends rose again during the Second World War.

The shale-oil industry in Scotland fell into further decline during the inter-war years, unable to compete with cheaper, imported crude oil, and operating at a loss by the early 1920s. The first spate of closures of oil works and of shale mines came in 1921–2 and further closures were accelerated by a six-week strike in the winter of 1925 against a 10 per cent wage reduction. For the shale-mining community, soup kitchens and hardship anticipated the experience of their coal-mining neighbours a few months later. The total workforce in shale-mining and the oil works had dwindled to around 4,000 by the mid 1930s, when only fifteen mines were operating, and miners were working short-time. In post-war decisions about fuel and energy needs, shale mining with its small (2,000) workforce was dismissed as an insignificant sector and was excluded from nationalisation in 1946.

Plate 59.

Shale miners, one using pick, the other drilling, preparatory to shot firing. (*West Lothian Council*)

Stoop and room working and hand hewn methods of coal extraction still persisted in all parts of the coalfield areas in 1939 (and continued as the norm in shale-mining), but the triumph of mechanised production in longwall operations had become most evident in Lanarkshire, where nearly 85 per cent of coal was machine cut, compared to the Scottish average of around 80 per cent. The Scottish coalfields had now become the most heavily mechanised region within the entire British mining industry; and it was also the most dangerous in which to work.

In Scotland, a depleted and weakened labour force of 80,000 mineworkers struggled to meet the demands of coal production during the Second World War. Many of the younger and most vigorous workers deserted the pits for the armed forces and for other, better-paid jobs and conditions in wartime. This drift away from mining was halted in 1941, when mining became a reserved occupation, and mineworkers were effectively conscripted as they were not allowed to leave their employment unless medically unfit or having reached retirement age. Government attempts to recruit men and boys into coal-mining employment were not successful, as pit work had a poor reputation and was not an attractive proposition, especially to young men who did not come from a mining environment. Despite the publicity surrounding it then and now, the compulsory recruitment of the 'Bevin Boys' was a contentious experiment that made little impact on fresh recruitment into the pits in Scotland and the typical mineworker in wartime had an older age profile than before 1939. Faced with the shortage of experienced and trained mineworkers, and under great pressure from wartime workplace regulation and coercion, Scottish miners worked overtime and otherwise made their own vital contribution to the war effort.

However, the long-term problems of a declining and failing coal-mining industry were fully exposed in wartime when Scottish output actually fell from 30.5 million tons in 1939 to 21.4 million tons in 1945. In 1944, the extent of these problems was starkly revealed in the report of a major wartime official enquiry into the condition and prospects of the Scottish coalfields. The members of the Scottish Coalfields' Commission were alarmed by the continuing decline of the once great Lanarkshire coalfield, where output had decreased dramatically during wartime. They acknowledged the exhaustion of most of the upper coal seams, the expectation and the expense of poor and thin yields from deeper seams and the short working-life prospects (5 to 20 years) of the bulk of its remaining 58 collieries. In their recommendations they could not justify state assistance to private owners to meet the enormous and prohibitive costs of de-watering and renewing the flooded mining districts of North Lanarkshire.

The four miners' union representatives on the Commission took a different view of prospects, advocating a fully planned, nationalised industry that would be accorded state resources to restore and develop collieries in the absence of inefficient and failed private ownership. However, there was general agreement that the economic decline of the Lanarkshire coalfield could not be reversed. They considered that, even with further investment, most existing collieries would remain unreliable as profitable units, and the development of new collieries in the area was also rejected for the same reason. Having passed this sober verdict on the future prospects of the Lanarkshire coalfield, they reckoned that, in the short and longer term, Scottish coal production and investment targets would have to concentrate on the more productive and economically viable parts of the Scottish coalfield. The Cumnock district of Ayrshire, Dumbarton and Stirling (but not Falkirk), Fife and Clackmannan and the Lothian counties were earmarked for further development. A few super pits, including the potential of seams under the Forth, were mooted as exciting prospects

However, the Commission had to admit that 'even the best-laid plans for the development of the Scottish coalfields must fail unless miners in sufficient numbers can be retained in, or attracted to, the industry. One essential condition of this is that an adequate number of suitable houses should be made available within reasonable access of the various collieries.' They calculated that a post-war programme of upwards of 30,000 new, improved-quality rented houses was required urgently to accommodate miners and their families:

> The present shortage of houses for miners – made up of houses needed to replace unfit houses, to end overcrowding, for miners who might reasonably live nearer their work, and for additional workers who could be absorbed at existing collieries – is about 19,000. In addition to making good the shortage above, the number of houses likely to be required for miners in connection with contemplated new sinkings is about 14,500. These should be erected as and when the new sinkings develop.[3]

Their proposal recommended the promise of additional and new accommodation for those mining families who could be encouraged to migrate from declining colliery areas to developing ones as, for instance, from Lanarkshire to Central Fife, where more work was becoming available at one or more of the projected giant pits.

The first stated priority was demolition of the remaining stock of older, unfit houses, particularly the scandalous, insanitary, overcrowded miners' rows and other sub-standard tied company houses that could not be renovated or converted. The Commission recognised that the large majority of older houses referred to so scathingly in the Report of the Royal Commission on Housing (1917) had long since been demolished,

> but unhappily there still remains a substantial number which can only be classed as 'unfit for human habitation'. These houses which, generally speaking, have already been condemned by the local authorities continue to be occupied only because of the extreme shortage and, but for the war, most of them would doubtless already have been demolished and replaced by new houses.[4]

They had inspected examples of the worst of the older type of miners' rows – at Auchinleck in Ayrshire, Glencraig and Lumphinnans in Fife, Blantyre in Lanarkshire, Niddrie in Midlothian and Cowie in Stirlingshire – and commented as follows:

> For drabness of appearance and atmosphere there is no grouping of dwelling houses we know which can compare with collections of the older miners' rows – dreary long lines of single-storey houses and between them only narrow strips of drying greens (though there is no grass left even if there ever was any) often with communal wash-houses, conveniences, and, in these times, air-raid shelters, scattered over the 'greens'. Sometimes the atmosphere is polluted with the noxious fumes from a burning bing nearby. Although in some districts these older rows are kept in quite good condition, in others their condition is deplorable and the houses are worse than damp. Another unfavourable feature of the older rows is that they consist usually of only two apartments. In view of the limited accommodation there is little room for families, and much overcrowding. In addition, we saw for ourselves cases where the coal had to be kept under the bed.[5]

In many mining areas, during the 1920s and 1930s, local authorities had taken the lead in building council houses for miners, although several coal companies had continued to build houses for their workers, particularly in the Midlothian area. Progressive local authorities had exercised their statutory powers to demolish the notorious miners' rows and overcrowded slums and to provide substantial programmes of state-assisted rented housing. Perhaps the best examples of this provision during the inter-war years, and continuing after 1945, was to be seen in Kilsyth, which formerly had some of the worst housing for miners. It was transformed into the town with the highest proportion of council housing in Scotland.[6] The large-scale introduction of council housing provision in mining communities was usually the result of successful political intervention, principally by Labour-controlled councils, aided by dedicated health and medical professionals. The Commission recognised the major responsibility of local authorities for public sector housing and looked forward to an expanding role for councils in providing improved accommodation and facilities for mineworkers, including well-heated, three- and four-bedroom houses, supplied with bathrooms and garden space. The Commission were also clear about signalling an end to tied company housing, especially in separate pit communities, as a practice which belonged to the past. They recommended that no more tied company houses should be built, as it deterred many workers from being recruited into the industry and into a mining area, and suggested that existing company stock which came up to modern standards be sold for owner occupation in the near future. Anticipating an era of post-war planning and regeneration, they stated a preference for substantial and mixed-occupation communities where a range of job opportunities outside of mining would be made available for members of mining families, and a residence pattern in which incoming mineworkers would be integrated with other workers. This joint proposal was also the position of the mining unions. However, there was an emphatic consensus that local authorities

Plate 60.
Arniston Colliery pit-head baths of unusual circular design, *c*.1933.
(*Collection of the Scottish Mining Museum Trust*)

should be the principal provider of accessible social amenities in each housing scheme, mining community and isolated pit village, working in partnership with agencies such as the Miners' Welfare Commission who were a source of major funds for this purpose.

The Miners' Welfare Fund had been created by government legislation in 1920, financed initially by a levy on coal output, and boosted in 1926 by a levy on coal royalties. The fund was intended to enhance social, recreational, health, welfare and educational facilities for mineworkers and their families, principally at community level, although assistance with housing was excluded from the scheme. Between 1921 and 1941, grants to a range of miners' welfare schemes in Scotland totalled £2.25 million (the total British coalfields' allocation was £21 million), and was divided into three main categories. Applications for funds and their allocation to the various amenities programmes were decided at local level by joint committees repre-senting colliery owners and mineworkers, and both of these bodies were also responsible for the management and maintenance of individual schemes.

In Scotland, over £1 million was spent on the provision of indoor and outdoor facilities for recreational and leisure pursuits. By 1941, nearly 300 schemes had benefited by grants from the fund. Typical individual and combined schemes included a park and recreation ground with pavilion, bowling green, children's playground and perhaps a swimming pool and a village football field. Other popular and widespread schemes included a well-furnished miners' institute building or community centre comprising rooms for a library and quiet reading space, table games, meetings and a larger hall for dances, community drama and functions. A small baths' or swimming-pool facility was sometimes provided at the larger institute complexes. In a pit village a simple building with rudimentary facilities made a difference in an otherwise barren environment.

Around £900,000 was spent on pit-head baths and associated pit-head welfare. Although progress in providing baths was very slow in the early 1920s, the programme took off after 1926, levelled off again during the slump years and rose steadily thereafter until it was suspended by the war. Provision was uneven across each coalfield area, and larger collieries with a projected long life were usually given priority. Altogether, within twenty years, eighty-two collieries were furnished with pit-head baths, giving accom-modation for 50,000 men (out of 80–90,000 in wartime) and around 300 women who worked at picking and sorting tables. To their credit some owners, such as the Coltness Iron Company at the new Kingshill Colliery complex in the late 1920s, continued to provide pit-head baths on their own account, but this example was rarely followed and provision was generally reliant on the welfare-fund source.[7] Other pit-head facilities eligible for the fund included clothes-drying and cloakrooms, colliery canteens, cycle-storage and boot-repair shops, although these items were not universally provided following the installation of baths and showers. The absence of pit-head welfare at many collieries was a bone of contention among miners, and even when facilities were provided, the miners had to pay the running costs of washing the dirt from their bodies and clothes. The miners' unions continued to press for comprehensive provision of pit-head baths and associated welfare facilities free of charge to the workers. While the welfare fund scheme made significant inroads, it was not until after nationalisation that the Coal Board was persuaded to implement a further programme of pit-head baths and only then to accept full responsibility for their full maintenance costs as a charge against the industry.

The third main category of welfare-fund provision comprised various health and convalescence services for miners and dependants. Most of the £200,000 spent in Scotland between the 1920s and early 1940s was devoted to the running costs of five convalescent homes. Accommodation in the west was provided for men at Saltcoats and Kirkmichael, and for women at Troon and Skelmorlie. In 1927, Blair Castle, Culross, originally the property of Fife Coal Company, was donated as a convalescent home. Whatton Lodge, at Gullane, was added in 1947 for the benefit of Lothian miners. As a general rule in these homes, respite was available for at least a fortnight or more for each person, including return visits, on the joint recommendation of the local miners' agent and doctor. Every year thousands of men and women from the mining communities gained from this service (in 1944, a combined total of 3,000 people enjoyed periods of residence in the five homes).[8]

In Fife, in addition to Blair Castle, the Miners' Union had its own convalescent home at Leven. This was the mansion of Charles Carlow, head of Fife Coal Company, who retired to St Andrews in 1947. His gift was turned into a splendid rest home for miners' wives.[9]

Despite its limited extent, the work of the welfare fund from the 1920s to the 1940s had a positive impact on the fabric and wellbeing of many mining communities, especially in rural pit villages. However, it was obvious that such provision could not eradicate or compensate largely for the social deprivation and decay of life as the direct consequences of low earnings and unemployment, inadequate housing and the lack of pit baths with laundry facilities. Although the best of mining villages had a varied and lively cultural life, this was essentially a man's world where women still had little or no scope for participation outside of homemaking and child rearing, and the daily grind of domestic chores, forever washing and cleaning the men's working clothes.

Working Conditions and Standards, 1920–1946

The Image O' God

Crawlin' aboot like a snail in the mud,
Covered wi'clammie blae,
Me, made after the image o'God –
Jings! but it's laughable, tae.

Howkin' awa' 'neath a mountain o' stane,
Gaspin' for want o' air,
The sweat makin' streams doon my bare back-bane,
And my knees a' hauckit and sair.

Strainin' and cursin' the hale shift through,
Half-starved, half-blin', half-mad,
And the gaffer he says, 'Less dirt in that coal
Or ye go up the pit, my lad!'

So I gi'e my life to the Nimmo squad,
For eicht and fower a day,
Me! Made after the image o' God –
Jings! But it's laughable, tae.[10]

The verse of Bowhill miner, socialist, poet and playwright, Joe Corrie, written in the 1920s, conveys something of the agonies, anger and frustration of the stripper and filler on a brutal shift, while feeding the mechanical conveyor at the coalface as a member of a contracted longwall team working under speeded-up and closely supervised conditions.

Other accounts of working conditions by miners elsewhere during the inter-war years and before nationalisation bear witness to intensified shift work and the increased risk of injury and debility in the highly pressurised environment of mechanised coal-cutting and moving conveyors. Such impressions are borne out in the findings of the Mines Inspectors' annual reports for Scotland from the mid 1920s to 1938. They report increases in the amount of serious accidents as compared to previous years, despite falling numbers of workers in a contracting industry. The higher levels of dust raised

by machine cutters and conveyors in the confined conditions of coalfaces and
haulage roads created more opportunities for lung disease, debility and
irritation. Increased dust levels also mixed dangerously with methane gas to
raise the risk of local explosions. Moreover, in wet places, some damaged
cables were liable to leak electric charge into the air, and when breathed in,
created an acid effect and taste. Between 1925 and 1935, the Scottish
coalfield had the highest incidence of accidents caused by electric shocks.

In his classic autobiography, Bob Smith recalled his early experience of
the mechanised conveyor on the 'Pan Run' at Ferniegair Colliery, near
Hamilton, in 1940. At twenty years old, he was then a proven hand hewer,
used to working on his side and his back in two-foot seams in stoop and
room sections. He noted the very different conditions in a mechanised
coalface, where increased dust levels and a much noisier working
environment added to health and safety anxieties and hazards:

> The conveyors were new to us. And yet another sign of the changing
> times in mining. They were troughs, each about seven feet long,
> mounted on rollers and overlapping. They were set on a slope along
> the length of the face, and were shaken back and forth lengthwise by

Plate 61.
Shale miners: one drilling,
the other shovelling on to
conveyor belt.
(*West Lothian Council*)

the Pan Engine. That made the coal travel down their length, and at the end it was loaded into the hutches. It was a deafening operation, and very dusty. None of the men liked it, even if it made the work somewhat easier. We had been used to the comparative quietness of working only with pick and shovel, so that we could hear the movement and creaking of the roof. We felt a lot safer when the place was quiet enough for that, because we judged the condition of the roof and the sides by the sounds they made. In the deafening din of the Pan Run, you could hear nothing else but the incessant roar of the machinery, and felt very defenceless.[11]

Working on machine runs at Hamilton Palace Colliery in the 1930s, David Meek was for a time a back-end man behind the cutter. He thoroughly detested this job as the worst he ever had as a mineworker. It was a job few workers would volunteer for, and few would keep for long if they had a chance of a move elsewhere, having to contend with the fumes, swallowing a lot of dust, suffering grit in the eyes without the benefit of masks, all the while trying to complete the task of clearing up after the machine and inserting timber props. The conditions were especially bad in a dry section, where flying dust was thick and wet-spraying to reduce dust levels was as yet non-existent.[12]

Digging coal in narrow seams in Lanarkshire's notorious wet pits, prone to flooding in deteriorating conditions during the inter-war period, was also to be avoided when possible. Waterproofs or oilskins were rarely issued. Bob Smith had more than his fair share of this work at Ferniegair, as he describes the scene at a fourteen-inch seam, where water had first to be laboriously bailed out into waiting hutches before the level was low enough to enable the men to get down to work, to take out the coal:

> The gum (small coal) from the bottom of the seam was wet, and stuck to the shovel as you tried to turn it out into the roadhead. As you lay on your side stripping coal, water was constantly dripping from the roof on top of you. It was ice-cold water and you felt pretty miserable. There were no oilskins or any other sort of protection: we simply worked in our trousers and singlets, lying in the wet mud and wielding shovels. We were paid an extra shilling a day for working wet.

On such occasions, 'it was a great relief when lousing time came. The men commented on our drookit state as we went to the pit bottom and gladly let us through to be first up the shaft and into the warmth of the showers and dry clothes.' This, at least, was one consolation, as Ferniegair was then (1940) one of the few pits in the area with baths and drying rooms. According to Smith, at this time, the nearest pit with baths in the large Larkhall-Hamilton coalfield was the Priory, at Blantyre, several miles away, indicating the gaps in provision of pit welfare. When he worked in wet, narrow seams in Shotts in the mid-1940s he was disgusted to find the complete absence of pit baths in this mining area with twenty pits. 'The results of too much wet work was rheumatics, beat elbows and knees, and boils and of course the work was much harder as the wet coal had to be handled in those cramped conditions. Constantly lying on your side in several inches of gritty mud, while trying to turn out coal into the roadhead was no joke.'[13]

While working in the Dalhousie section of the modern Lady Victoria Colliery at Newtongrange, James Darling reminds us that wet conditions were not confined to Lanarkshire:

> Right in the middle o' the coal face the water wis two feet deep. Now ye can imagine a seam three and a half feet high, and the water aboot two feet. And you're supposed tae go in and clean that coal off. It wis cut wi' a machine, ye see. But ye crawled up in the conveyors tae keep yersel' dry. Ye stepped oot the conveyor, right up tae the waist in water. And ye worked in that the whole day, up tae the waist. So the pay day came, and of course production wis doon, wages wis doon tae under seeven bob a day.[14]

Darling, among others, refused to work in such conditions for dirt wages, and walked out of the job. That infamous manager, Mungo Mackay, kept him idle for three months, to teach him a lesson for causing trouble. With a surplus of miners looking for work, managers were in a strong position to dictate terms, to choose and reject their workers. Tommy Kerr, working at the Ormiston pits, in East Lothian, also discovered that complaining and agitating about low wages and wet conditions in the early 1930s could result in being put back into similar conditions by vindictive managers and contractors: 'In these days ee didnae win ee're place at the coal face wi' a draw oot the hat for a place: ee were put in the place. And if you were working wi' a contractor and the contractor got this information frae them that employed him, ee was put in the worst place – the yin wi' water, the yin maist difficult tae work.'[15]

Bob Smith's account of his life as a miner during this period returns again

and again to the dangers, the constant injury toll and the human cost of coal-mining. Leaving the wet, narrow places and insidious blackdamp of Ferniegair for the deep, dry, hot, gassy Bothwell Castle Colliery, with its steep inclined haulage roads, high main roads (but low secondary roads with frequent roof collapses) and seven-foot seams, brought a new mix of hazards and possibilities.

Once again he was a pick-and-shovel collier, working his own place, holing and shot-firing, but having to test frequently for gas, put up screens to aid ventilation and to do repairs to the immediate working environment. He especially resented having to do his own haulage, as he and a mate had to heave their loaded hutches manually up the incline. 'The roads were getting longer as we worked the seams, and we had much further to draw our hutches. I hated that part of the job, regarding it as cuddy's work. The roads were long and some of them steep. There was no wonder that so many men were suffering from strained backs, broken legs and so on. It was brutally hard work.' Later on, shifting to a place above the level, they had to push the empty hutches up, and take the full ones down the incline. This made for easier but more dangerous work:

Plate 62.

Steep working seam at Aitken colliery, Fife. (*Collection of the Scottish Mining Museum Trust*)

Those full hutches were heavy, and we had to keep them under control going down the steep slope. We had various ways of slowing them down. Sometimes we spread screen cloth on parts of the road, and then used double snibbles or snags on the wheels, with sand on the rails to give the locked wheels a better grip. We had to hang onto the handles at the back of the hutch, and dig our heels into the roadway between the sleepers, trying to control it. Even so, with everything we could do, there were occasions when a hutch ran away with us, and then we would race along behind it, shouting as loud as we could 'Runaway! Runaway!' and race along, hoping it would not jump the rails as the hutch sped down the track at a tremendous speed. If it did jump the rails, there was a wreck, and at the very least we had to labour hard to get it back, and at the worst it could hit someone, or else knock out props and cause a fall.

In his short six months at Bothwell Castle Colliery, Smith survived a stray shot-firing explosion in which slivers of stone and coal heavily lacerated his back and arms, and he suffered a bad injury to his arm in a crushing incident with a hutch. He also assisted at a runaway hutch incident involving the young footballer Jock Stein who had been newly transferred to Albion Rovers from his local junior club, Blantyre Victoria. Stein's leg was jammed between a hutch and the motor-driven haulage winch he was handling. 'The way was almost blocked, and only the smallest of us, like me, could squeeze through. We found the roadsman beside the agonised Jock, and other men rushing up from the other side. Between us, we managed to lift the hutch, stell it up, and pull Jock clear. The fireman arrived, and ordered a stretcher party to take him up the pit, where it was found that his leg was badly injured.' The determined Stein recovered and, as is well known, subsequently became a legend in the world of football.

Bob Smith witnessed several other serious incidents there, including horrifying injuries to a young drawer, who lost a foot and an eye. 'Sometimes it seemed that the coal we took out of the Bothwell Castle demanded a constant sacrifice of blood,' and he was determined to get out of a dangerous, accident-prone pit where seven lives were lost in several incidents between the end of 1942 and early 1943.[16]

The annual Mines Inspectors' reports for Scotland expressed frustration at the amount of avoidable accidents due to failure to secure roofs by adequate propping, indicting both workers and managers alike for a lack of care. While some miners did take risks by cutting too far into a coalface without using supports, they often did so under pressure to make a meagre living from poor piecework rates, and in many cases sufficient timber and props were not readily available or were in short supply. Even in new sections of collieries, the official reports often criticised managers for deficient road and roof maintenance, for poor ventilation and for concentrating more on production at the expense of safety. Such deficiencies were persistent in older and smaller pits where miners were vulnerable to roof falls and gas poisoning. While powerful fans continued to be installed in the modern and larger pits, environmental conditions in many of the other pits remained poor or were worsened by owner or managerial cost-cutting and neglect of regulations. Although an extreme example, it is almost incredible that the most primitive, age-old, makeshift method of dealing with gas was still being relied on in some pits, as in this oral testimony from Ayrshire between the wars, where a couple of older, unfit ex-miners were, 'down the pit on shift work, wafting screen cloth around to clear the gas away from the coalface before the men went in. ... They were birlin this screen cloth round about their head, and going in a wee bit further and daen the same.'[17]

The major pit disasters of the inter-war and wartime period, involving large-scale loss of life, were caused mainly by explosions, as at Plean (1922); Gartshore, Twechar (1923); Bowhill (1931); and the worst, at Valleyfield, Culross (1939) when thirty-five miners were killed. Following this disaster, management were prosecuted and fined for breach of safety regulations, including failure to control use and storage of explosives for shot-firing. An electrical fire claimed nine lives at Dumbreck Colliery, Kilsyth, in 1938, but the greatest single loss of life was in 1923 at Redding, near Falkirk, when

Plate 63.
Valleyfield Colliery disaster, 1939: awaiting news outside the gates.
(*Collection of the Scottish Mining Museum Trust*)

forty were drowned and suffocated by an inrush of water and earth.

On the positive side, during the inter-war period, there was some official recognition of the need to provide recuperation and rehabilitation facilities for injured and disabled miners. When as many as 7,000 miners were being injured every year in Lanarkshire alone in the early 1930s, a dire need existed for a range of occupational-health clinics and rehabilitation centres. The first orthopaedic outpatient clinics for miners in Britain were opened in Lanarkshire in 1935, set up in miners' institutes and public-health centres. They were aimed at helping miners to recover from injury and restore their fitness for work. However, while residential rehabilitation centres had long been promoted by the mining unions, only one centre was eventually provided at any time prior to the National Health Service. It was located at Gleneagles Hotel in 1943 and, after closure in 1945, it was succeeded by a dedicated day centre established at Uddingston, Lanarkshire, in the former miners' institute building, under the auspices of the Miners' Welfare Commission. The enlarged building and grounds had extensive sports and recreation facilities, workshops and a library, and were staffed with surgeons, physiotherapists, social workers and fitness trainers. A similar centre was planned for the eastern coalfield, located in Fife, but was never built in this period, and reflects failed priorities when there was an obvious case for more centres of this kind.[18]

Conflict and Community, 1921–26

The General Strike

As lads we ran aboot the braes
In wee bare feet an' ragged claes;
Nae such a thing as 'Dinna Like',
For then oor faithers were on strike.

Yet in these times they still could sing
While haulin' hoose coal frae the bing;
Nothing then tae waste or spare,
Still everyone would get their share.

They'd share their last with those in need,
There wisnae such a thing as greed,
A piece on jam was something rare,
An' no so much o' that tae spare.

FIT AGAIN

Rehabilitation for Injured Miners
Uddingston Out-Patients' Centre
MINERS WELFARE COMMISSION

We'd eat it new, we'd eat it stale,
And even dip it in oor kail;
Nae such a thing as 'Dinna Like',
We kent oor faithers were on strike.[19]

The often traumatic experiences of the strikes and lockouts of 1921 and 1926 and the mental and physical struggles to make ends meet in the depression years of mass unemployment, short-time working and poverty wages, were seared into the minds of mining communities. In Scotland, this popular memory of hardship, solidarity and struggle continued to resonate, especially in later battles, as in 1972, 1974 and 1984–5, and still refuses to die into the new millennium among survivors of mining communities.

The 1921 Lockout

In the three-month lockout of 1921, the principal storm centres of direct action and communal resistance were in parts of Fife and the Lanarkshire coalfield. Although Ayrshire and the Lothians were relatively quiet, they also included some incidents of sporadic but determined picketing, which spilled over into violence and clashes with police. As we have seen, places such as Bowhill, Lumphinnans, Cowdenbeath, Blantyre and Shotts had an established tradition of industrial militancy. There, younger miners in particular were also strongly influenced by Marxist revolutionary ideas. Their political activism was directed against private ownership of the industry and the war-mongering tendencies of the capitalist system. Organised in the Miners' Reform Movement, they were also impatient at the moderate stance of the old-guard leaders in the mining unions, and campaigned to replace them with radical policies and militant leadership. However, in 1921, despite these differences, miners of all persuasions were outraged by the failure of the owners and the Government to recognise the justice of their claims as wage earners and vital producers of wealth.

As John McArthur explained in his recollection of the 1921 episode:

The miners never accepted 1921 as a strike; they always claimed that it was a lockout. The Government had de-controlled the mines just beforehand, and the owners issued an ultimatum that the mines would only continue provided reductions in wages were accepted by the miners. In this way they also gave notice to the safety workers who were employed in the pits. For the first time in the knowledge of the trade union the miners and safety workers took joint action.

Plate 64.
Rehabilitation poster.

Every one in the pits was withdrawn – tradesmen, safety workers, pumpmen, winding enginemen joined the miners who were locked out. This created quite a furore among the employers, the Government and the Press. They said we were completely damaging the industry for all time, that the pits would be flooded because pumping workmen were withdrawn. They said we were deliberately stopping the pumps and allowing water to creep up … In Fife, as elsewhere, the coalowners tried to keep pumps going by recruiting scab labour. They had conducted a campaign prior to the lockout, and they were able to man a number of pits with misguided students and colliery officials and clerks. Before very long, however, the strongest opposition was evidenced amongst the miners over this. In almost every part of Fife, spontaneous mass meetings of the men almost simultaneously objected to the steam-raising and pumping being continued by the blacklegs. The mass meetings decided to march on the pits and demand the withdrawal of these men.[20]

Union policy to stop the pits completely by resolute picketing was a calculated tactic to bring the lockout to a speedy end and force the owners to reopen the pits and wages negotiations. For example, in central Fife, undeterred by police detachments drafted in to guard the mines, direct action at times took the form of a communal picket. Crowds of several thousands, including women, systematically toured the pits in their area to ensure that power supplies were stopped and nobody was working. Thus, Bowhill and

Plate 65.

Left, John McArthur.

Plate 66.

Right, Abe Moffat.

Glencraig were forcibly closed on day one. Two days later, around 2,000 miners headed by a piper compelled stoppage of pumping at Leven and Wellesley collieries, while an even larger crowd, fronted by union officials and Cowdenbeath and Hill of Beath pipe bands, marched to Kirkford Colliery to close the pits. Blacklegs were manhandled and the unpopular manager at Dalbeath was frogmarched through the streets of Cowdenbeath. Police had to mount four baton charges before they finally rescued the manager, made arrests and dispersed the crowds.[21]

Similar scenes of flying, mass picketing to ensure cessation of pumping were often enacted successfully in north Lanarkshire, as at Shotts and Harthill, though not without clashes and provocation between pickets and police, and occasional harassment of blacklegs. In Midlothian, renowned for its moderate trade union tradition, the area miners' leader, Andrew Clarke, denounced the decision of his own union to withdraw all maintenance workers. Nevertheless, the managerial regime of Mungo Mackay at Newtongrange was temporarily breached by mass picketing. A huge procession of miners and families drawn from Dalkeith and surrounding areas, the men armed with staves, marched into Newtongrange to the

Plate 67.

Soldiers guarding a Fife pit during 1921 lockout. (*Collection of the Scottish Mining Museum Trust*)

company offices and effectively ordered the stoppage of boiler fires at the Lady Victoria and neighbouring collieries, including Arniston. The Highland Light Infantry arrived the following day to guard the pits and protect the blacklegs who stayed inside the premises and were reputedly fed and treated like lords throughout the lockout.[22]

In Ayrshire, the official union leadership counselled against preventing collieries from organising their own maintenance operations, a decision largely obeyed by the miners. However, demonstrations and mass pickets forced stoppage of pumping at pits in Glenbuck, Muirkirk, Dalmellington, Patna and Cumnock, although flying pickets from Dalmellington were repulsed at Bank Colliery, New Cumnock, by a company man with a revolver and high-powered hose-pipes.[23]

The early phase of disorder had prompted the authorities to rush police and military reinforcements to immediate trouble spots, particularly in central Fife where they were billeted in mining towns and villages. Here, as in other parts of the coalfields, their presence was usually sufficient to prevent further intrusions by pickets. Public sympathy for the case of the miners was well known, and in Fife there are strong hints of some Scottish troops being inclined to fraternise with the miners although they, and detachments of marines, carried out their essential duties of guarding collieries.

As a leading militant, John McArthur admitted that it was impossible to exert control and maintain discipline while handling thousands of angry and frustrated miners mobilised for direct action, and excessive behaviour could be understood, if not entirely condoned. Convictions for mobbing, sabotage, assault and attacks on police or retaliation against them carried heavy prison sentences, and many militants paid the price for their actions. Hooligan elements engaging in selfish looting and other senseless criminal behaviour were generally condemned within the working class and beyond. Random violence was rare and militant behaviour, in any case, whether legal or illegal, was always going to be attacked and made into a sensation in certain quarters. At the end of April, the Tory MP for North Lanark, Mr R. McLaren, declared in the House of Commons that the wreckers had won in his own district, as many mines were already ruined by flooding. He was convinced that the strike was the work of Communist agitators and extremists, not decent, law-abiding miners.[24] However, McLaren was wrong to search for a 'red conspiracy' behind the aggressive conduct of the miners during the dispute. Certainly, among the younger generation of miners were industrial militants and revolutionary socialists such as John McArthur, who had joined the Communist Party (formed in 1920–1), or belonged to no party, and who had a high profile as political and union activists in their localities. They sought to give a lead and did command some influence but, throughout April to June, the impact of a few Communist militants in any locality could not be responsible for the manner in which almost the whole mining community in Scotland stood firm. At large meetings, miners everywhere reaffirmed their determination to refuse to submit to the harsh terms tabled by the coal-owners, and made this known to the national executive of their union.

However, to dwell solely on the many open confrontations described above is to give a misleading impression of the everyday experience of the lockout within mining communities. Throughout the dispute, the main preoccupation was to keep body and soul together, and most energy was expended on soup kitchens and other communal programmes to sustain solidarity and morale. Union funds were inadequate to support generous and sustained payments to mining families, and were not intended for use on this scale during a prolonged period without earnings.

Due to the emergency created by the lockout, no fewer than half a million Scots within mining communities were without income for three months or more, as the Scottish Board of Health would not allow relief payments to unemployed workers who were able-bodied. Those miners who had paid into their national insurance scheme suffered an early setback when they were refused out-of-work benefit under the Unemployment Insurance Act, as they were classified as strikers or as locked-out workers. Moreover, already debarred from parish relief, only their families and dependants could claim allowances from the parish under the strict and humiliating terms of the Scottish poor-law system, and such claims relied upon lodging successful applications on the grounds of proven destitution and good character. Obviously, there were differing views within the mining community and within families about submitting to the indignities of parish relief, and while many eventually decided they had to go down this route, many others chose stubbornly to retain their pride and independence rather than submit to a system they despised.

Soup kitchens, mainly organised by union members and staffed by miners' wives, were set up for communal feeding, and adequate stocks of provisions were sought. As John McArthur explained of this effort in Fife:

Because of the shortage of union funds, it was necessary to make sure that the miners were not starved into submission. The cheapest and most effective way to feed masses of people with a minimum of money was to organise central kitchens, which would be run by the wives with the help of the miners themselves. The kitchens would cook on a mass basis. So these kitchens were opened up in each locality, controlled by the miners' strike committee. Strenuous efforts were made to obtain bulk food supplies, rather than distribute individual donations as strike pay. This meant that you could prevent actual starvation from forcing a return to work.[25]

The concession that had been won earlier from school boards (local educational authorities from 1919) to provide modest free meals for the school children of miners during prolonged industrial stoppages was again implemented, and was extended into the weekends. At Cardenden, Fife, a former pupil recalled how, in 1921,

> I used to get my dinner and that at the school. You went down and got a roll in the morning, a large roll, what you called a penny roll. Then you got two slices of bread and a bowl of soup at dinner time, in the school playground. It was made at the school. When you came out of the school – two slices of bread and jam. And on a Sunday you got a boiled egg, your mug of tea and a slice of bread or toast.[26]

As well as contributions of food from local farmers, shopkeepers and other sympathisers, many of the men went out poaching for rabbits and fish. Poultry and sheep also went missing from some farms, although miners generally observed a code of honour whereby friendly and helpful farmers were not harassed. In addition to centralised kitchens, soup and meat were often cooked in the scrubbed-out boilers in the scullery of a miner's house. Mary Docherty, daughter of a miner and activist, tells how

> the police came to our door and asked my mother if anyone had been trying to sell her beef or had given her beef. My mother said she had no money to buy beef and nobody had given her any. The constable was kept at the door but he could see into the kitchen, saw a pot boiling and remarked on it.
>
> 'Do you know your pot's boiling, Mrs Docherty?
>
> 'Aye, I want it to boil. That's why I put it on the fire.'
>
> 'You're surely going to have a feed the day.'
>
> 'Well, it'll be a change from soup.'
>
> He tried every way to get to know what was in the pot but my mother did not tell him.
>
> About an hour or so later he came back to the door. My mother said to him: 'You want to know what's in the pot. It's a hen,' and she

lifted the lid of the pot so he could see and smell it. She thought it better to let him know in case he thought my father had begun helping himself to one of the farmer's sheep.[27]

Above all, the co-operative societies proved the mainstay of material support to mining families and were often their saviour in the long weeks without earnings. For instance, in Shotts, the Co-operative Society had 2,300 members in 1921, representing almost every mining family. The Co-op was well placed to provide credit, and responded magnificently in the interests of its members and customers. From the end of the nineteenth century, and especially since 1914, the co-op movement had grown and, in mining communities, its members had shown where loyalties lay by banking and investing when they could, taking out deposit accounts and share capital. As customers at the shop counters, buying from a wide range of co-op goods and services, they earned a dividend on their spending as a form of savings or as discount. During the 1921 emergency, members were allowed to withdraw their savings and some share capital, to cope with hardships arising from the sudden loss of wages coming into the house. The stores gave further concessions on the price of bread and groceries. Understandably, the confectionery side of the bakery lost a great amount of trade, as did drapery departments, with the drastic fall in demand for clothes. Nevertheless, the Co-op in Shotts and elsewhere also donated large amounts of food for the soup kitchens. It survived as a viable body by paying out no dividends in the autumn of 1921, and by the promise of eventual recovery of debt from a grateful and loyal membership.[28] Where substantial co-operative societies existed in mining areas, their role was vital in keeping the wolf from the door.

While the dispute lasted, miners were without their usual concession of domestic coal, but this shortage of fuel for both home use and the soup kitchen boilers was soon overcome. Had it been the depths of winter instead of early summer, the problem would have been more acute but, at any time of year, there were always homes stricken with illness, and where elderly people and young children needed the warmth that only coal fires could then supply. At times, keeping away from the watchful eyes of police and landowners, miners got to work on old, disused shallow seams or on outcrops. This could be a dangerous and illegal practice, and was forbidden by the mining trade unions, although this ruling was relaxed in such extreme circumstances, and precautions were usually taken to ensure such workings were safe. Moreover, it was again a common practice to see men, women and children howking and scavenging small bits of coal from waste bings, keeping a lookout if permission had not been granted.

However hard hit, mining communities did not allow their difficulties to put a temporary end to seasonal festivities as, for example, in Shotts, the third annual Miners' Gala Day was held at the beginning of May in fine weather. A well-attended sports day on the Saturday was resumed the following Wednesday in a farmer's field, and musical displays were provided by the silver band and the Dykehead and Shotts Caledonian Pipe Band, both parading the town. The bands, concert parties and talented musical individuals alike among the miners were a familiar sight within their own communities. Jimmy Shand, from East Wemyss, legendary melodeon player and future Scottish dance-band leader, launched his career during the 1926 lockout as one of the miner musicians who toured nearby towns and much further afield, to raise funds for the soup kitchens. However, neither the activists among the pickets, the soup kitchens and the fund-raisers, nor the thousands they helped were to know that all these efforts would have to be renewed with even greater commitment, during the even more prolonged and drastic crisis of the 1926 lockout.

Strike and Lockout: May–November 1926

Since September 1921, long-term unemployed adults in Scotland who were facing destitution were eligible to apply for public assistance from the poor law, but strikers and locked-out workers were still not legally entitled to such provision in 1926, although their families were. During the emergency, the parish councils, as administrators of poor relief, were instructed to pay no more than the maximum rate of benefits to claimants, which was twelve shillings for a wife and four shillings for each child per week. This represented around a third of a miner's average weekly wage and was below bare subsistence level. However, the political complexion of parish councils in mining areas varied from place to place, and their attitudes ranged from outright hostility to generous support to claimants. In this respect, the situation was different from 1921 as, since then, more Labour, Socialist and Communist activists, including significant numbers of miners, had been elected to parish councils since 1922 and in some cases formed a majority. Several of these councils were prepared to defy the letter of the law and the government rulings about levels of assistance to out-of-work miners and dependants. Other councils that were hostile or wavering in this regard found themselves liable to mass pressure from the mining community and other working-class support, to grant at least the legal minimum rate for the duration of the dispute.[29]

In central and east Fife, the close-knit mining communities were then

Plate 68.

Soup-kitchen scene, Camp Street, Motherwell, 1926. (*North Lanarkshire Council*)

among the best organised and most militant in Britain. Wemyss Parish Council, dominated by miners' representatives, typified the local public bodies rallying to defend and support their own people. Thirteen out of fifteen Labour and Communist candidates had been elected in 1925, the other two members representing the coal company. During the 1926 crisis, this new, politically militant council felt they had the necessary power-base to exceed the limits of their statutory powers and responsibilities in the interests of the needy in this predominantly mining district. The council campaigned to get all miners to apply for public assistance as a right, and this was successfully done, as John McArthur, then a Communist member of the parish council, explained:

> This meant continuous meetings and whipping up round every street in the area of the parish council. We had to get the maximum number of people to apply for relief. This was not easy because a number of people were afraid of what their position would be legally if they had their own house, such as many of the miner-fishermen had. Some had money in the Co-op, some had other members of the family working. All these difficulties had to be overcome. And it was quite a sight to see this queue lined up at the parish council, including a number of well-dressed, respectable miners, applying for public assistance. This showed they accepted leadership on this question [instead of regarding themselves as charity cases].

More so than in 1921, communal feeding of adults and children was systematically organised, with the parish council paying for supplies, placing orders with the Co-op and welcoming any donations of food from other sources. As the weeks wore on, council funds were dwindling fast, so the council then decided to force the crucial political issue of guaranteed public maintenance for locked-out miners, especially for single men and pit-head women who were not recognised as legitimate claimants:

> And that to show what our entitlement was, we would have a mass demonstration to the nearest poorhouse, and for the males to demand admission. This was on the understanding that we could show up how impossible it was, and that the alternative to admission to the poorhouse was the continuation of payment by the parish council. The campaign was organised on a wide scale, but in our area it meant we had to organise the whole of east Fife for a march on the poorhouse which was situated in Thornton ... We were concerned to

make it an absolutely disciplined, organised march, and in the early part of the day instead of at night. We ran a series of meetings to explain the purpose of the demonstration and how we wanted this demonstration to be under the control of the Workers' Defence Corps that had been set up during the General Strike and had continued in being.

Under strong police surveillance, around 3,000 male and female marchers arrived in Thornton in orderly, disciplined fashion, 'and a deputation went to the poorhouse to demand admission ... the result of the demonstration was that the poorhouse said they had no room for us. The Government did not order the Parish to stop relief.'[30]

Mary Docherty recalled a second mass protest march held on the same day as the Methil to Thornton march (20 June). She participated in the memorable communal trek from Cowdenbeath to Dunfermline, several miles away. A huge procession of nearly 6,000 miners: wives, single women and children finally converged on Dunfermline Public Park, 'where a deputation was elected to go to the governor of the poor house to ask for the women and children to be taken in or granted outdoor relief. This was granted. There were too many for the size of the poor house.' Mary and her mother had marched to the end of Cowdenbeath before deciding to go the whole way. Mary was proud of the occasion, remembering her skinned heels and sore feet, and the kindness of the nameless person who gave them their tram fare home.[31]

Determined local leadership and communal direct action brought about similar victories in several other mining communities. Ballingry parish council (which included Cowdenbeath and 'Little Moscow' Lumphinnans), where Abe Moffat was a young Communist councillor, paid out-of-work allowances to miners as a right. Later, each councillor was surcharged £172 for making excessive and illegal payments, although Moffat claims that 'we refused to pay any of the money back, as we considered our action to be right in the circumstances.'[32]

Even in the heart of the Midlothian coalfield, at Newtongrange, a recent political landslide of ten out of eleven Labour candidates elected to the parish council helped to ensure the continuing welfare of locked-out miners and their families. Despite the usual workplace domination exercised by Mungo Mackay and his managers, their influence did not extend to control of voting behaviour. Miners' wives had a parish allowance of ten shillings a week, plus four shillings for one child, and three shillings for every other child. Communal soup kitchens and free feeding of school children also applied.

Unlike miners in Lanarkshire and Fife, miners in Midlothian, East Lothian and Ayrshire had a cushion of financial support from their trade union, as their larger funds supported a much smaller membership. Yet, by June, those funds were exhausted, and members had to fall back on parish council allowances, wherever available. In May, the mining unions in Soviet Russia had donated around £1 million to aid the striking British miners. This money was dispensed throughout the coalfields, with £28,000, the first instalment of the Lanarkshire share, being received at the end of May. Although a large sum of money, it obviously could not go very far among 60,000 men and families in one coalfield, but '2s.6d [12.5p] from Russia money' helped out single men who were not eligible for parish funds.[33]

The policy adopted by parish councils at Shotts and Cambusnethan was typical of mixed mining and industrial districts where Labour and left-wing members did not command a majority. By the end of May, Shotts Parish Council had admitted 1,500 claims on the grounds of destitution, including over 4,800 dependants. Shotts paid out the regulation amount of allowance to applicants among women and children, but chose not to make cash payments. This council displayed the Victorian attitude of contempt for recipients of poor relief. In their eyes, striking miners and their families were irresponsible people, and money payments were regarded as inappropriate, likely to encourage drunkenness and other reckless, feckless spending. Instead, the allowances were given out in the form of food vouchers which could be exchanged at local stores.

Although the rates level in Shotts was only fourpence in the pound, the longer the lockout lasted, councils became all the more concerned to economise on this large-scale disbursement of ratepayers' money. In July, Shotts cut the value of the weekly vouchers to eleven shillings for an adult and three shillings for a child. The adjacent Cambusnethan Parish Council was even meaner, and mining families suffered the consequences in September, when relief payments were stopped.[34]

In rural areas such as the upper ward of Lanarkshire, where mining communities were small, isolated and scattered across large parish districts, and councils were slow or opposed to support the miners by legal means, resources were harder to find in such times of hardship. Guy Bolton, then a young miner from Coalburn, acted as a hunter and gatherer in an unofficial quartermaster role during the long 1926 layoff:

And of course it was the same then as it was in the '21 strike. Ye had tae go oot and look for your food because there were nothing, nothing. And we went and we got bolder. It was rabbits and hens.

We always kept dogs for the poachin', no racin' dugs but poachin' dugs that could kill a hare or rabbits. We thought we'd take a change tae mutton, and the smell, the smell o' mutton cookin'! But unfortunately we filled too many pots, and so many o' us got arrested and ta'en away to Lanark. There were three or fower teams o' brothers. So they took a decision that the aulder brothers would take the punishment and the younger brothers would get off. And I mind the judge tellin' them: 'Ye know, I could sentence ye to death, to be hung by the neck until ye're dead for this offence. That's the law of the land for stealin' sheep.'

Thus his older brother Jock Bolton was one of five miners sentenced to three months' hard labour in Barlinnie for stealing and slaughtering sheep. 'When they came oot the jail, the silver band met them at the station. There was a welcome home party. They were seen as martyrs.'[35]

In larger clusters of mining communities angry but disciplined protest demonstrations were organised into the autumn against parish authorities that cut or withdrew welfare payments. Mobilising women and children in marches and deputations to local-authority offices was an effective tactic. One of those episodes has gone down in folk memory in the shale and coal-mining community of West Lothian. In West Calder, at the end of August, Sarah Moore (affectionately called 'Ma Moore'), Labour councillor for Addiewell, led a sit-down protest over several days against the decision to cease payments to families of locked-out workers. Mrs Moore had summoned claimants from the surrounding villages by bell-ringers, to march en masse to West Calder to demand their rights. Arrangements for food were made, and women and girls took their turn to camp out in front of the parish chambers in a non-violent protest. On the Monday morning, a few male protesters had joined the swelling numbers of mothers and toddlers, and the police were called in. They discovered a good-natured crowd but insisted on forcing them into orderly queues. When a man was pushed to the pavement, his bleeding head and heavy-handed police treatment provoked anger. The police resorted to batons to restore order as a couple of vehicles were overturned. However, concerned about public safety, the council conferred with Moore as the spokesperson of the demonstrators and immediately decided to restore relief payments on the following day.[36]

The locked-out miners remained remarkably solid throughout the first four months, through a warm summer when they saw more sun and sky than since their boyhood days, and family mouths were reasonably fed. There was time for outings and festivities, with concerts, galas, sports and games.

However, from September the weather and the predominant mood changed and hardship set in, as parish council allowances were cut or withdrawn, food vouchers were rationed or ran out and there was little or no money left for clothing, footwear, coal and rent payments. With no settlement in sight, doubt and demoralisation increasingly took over and hastened the trickle of miners returning to work under police protection.

In this final phase of the lockout, confrontations between the authorities and locked-out activists became more prevalent, as militants who resorted to desperate measures to avert the drift back to work increasingly provoked police retaliation. Unlike 1921, official union policy had instructed safety men to carry out maintenance operations as long as companies did not produce or move out coal, but militants at this late stage sought to intensify the struggle by picketing the safety men and aggressive harassment of others who were going back. From the beginning of the strike and lockout, police deployment had been more systematically organised and concentrated than in 1921, while troops were kept in readiness away from the front line. The Government had prepared its forces, including recruitment of ex-army men as special constables. As a paramilitary force stationed in the main trouble centres of Lanarkshire and Fife, these emergency constables had a notorious reputation for brutal and indiscriminate violent behaviour.

Most mining communities without a heavy police presence remained tense but relatively free from open conflict. However, in a few localities, mutual antagonism had resulted in ugly periodic running battles and clashes between miners and police. In the militant mining communities in and around Cowdenbeath, the area was under virtual martial law by September, after determined picketing to prevent movement of coal at Glencraig, and Bowhill had been countered by baton charges and mass arrests. One night, in Glencraig village, the police had run amok in the streets and among the rows, chasing and batoning everyone in sight, including women and youngsters, as well as miners. Throughout the district, public and protest meetings were banned and speakers who defied the order, among them prominent Communist and Labour activists and elected councillors, were arrested and convicted for incitement and sedition. Meanwhile, companies were evicting miners from their houses and installing blacklegs in their place. It is little wonder that such provocations prompted acts of resistance and retaliation by miner activists, to the extent of attacking busloads of police, dynamiting house windows of blacklegs, and sabotaging colliery property. Towards the end of the lockout, Glencraig activists broke into the colliery and sent cages down the shaft, causing a lot of damage. The police took their revenge, suspects being dragged from their beds in the middle of the night after their doors had been beaten in. The punishing outcome was twenty convictions and several long prison sentences.[37]

By the official end of the dispute in November, and surrender on the employers' terms, many thousands of miners had already gone back to work. Even at the bitter end, some militant communities remained solidly against surrender. Methil, for instance, was still superbly organised by leading local Communist activist David Proudfoot, and providing three meals every day at its communal canteens, subsidised by a small weekly levy on every household. However, already across the Scottish coalfields, victimisation was on a much larger scale than ever before, as hundreds of militants, union activists and many other miners who had publicly taken a principled stand were sacked and blacklisted from the pits and from other employment for years to come.

In the immediate aftermath of the national dispute, closures, rundown and restructuring of pits led to further unemployment in the most vulnerable mining communities. In those areas, the health and wellbeing, especially of women and young children, continued to be sacrificed as meagre allowances from parish relief or the labour exchange failed to stem privation. Yet, despite grim defeats on the industrial front and lower living standards, miners and their families stayed remarkably resilient in adversity during the inter-war years.

Beyond 1926: Conflict and Community

In 1927, a former leading coalfield and Communist activist made a perceptive assessment of political and cultural change within the Scottish mining communities he knew so well. He noted the prominent influence of Communist Party activists in Lanarkshire and Fife, and their recent election to leadership of the mining unions at county level. The swing to the political left among miners in the 1920s was also reflected more generally in the election at all levels from parish to Parliament of many Labour candidates who identified closely with the cause of the miners and their communities.

> In the tremendous struggle waged in the coalfields last year, the Communists took the lead everywhere. It was they who organised and ran the soup kitchens in the areas where parish relief had been stopped; they who arranged the concerts and meetings that kept the masses in tune. In every conflict with the law regarding picketing there were sure to be some members among the arrested. Certainly, the extremists may bulk more largely in the union's affairs at the moment,

because the membership is down to bedrock. Conservative and Liberal working men might incline to leave the union after a defeat; the creed of the Socialist or Communist forbids him this. But it is not really open to any doubt that the progressive decline in the standard of living of the miners since 1921, the series of titanic struggles waged, and the apparent hopelessness of any improvement along ordinary lines, has disturbed if not totally changed the political convictions of the average collier. It is not more than a single generation since the great mass of the miners in Lanarkshire and Fifeshire were Liberals; today there is not a single Liberal sits for a Scottish mining seat, and there are already several constituencies where the sitting Labour member could be ousted by a concentrated Communist attack.

It is true that more than the political ideas prevailing in the coalfields have been changed. A silent revolution in manners and customs has been effected which seems to pass almost unnoticed. The younger generation of miners is much soberer than the old. The bonnets and mufflers that used to be a respectable week-day attire have been discarded for hats and collars. Where their fathers used to slumber contentedly in the traditional mist of Calvinist or Catholic dogma, the new race of colliers, having been bitten with the worm of scepticism, are wide awake intellectually. They have a genuine thirst for knowledge. The extreme parochialism of the 'backwoodsmen' in the isolated mining villages and hamlets has vanished, retreated hastily before the conquering hoot of the motor omnibus, the herald of picture houses, fried fish shops, and sporting papers. The semi-rural character of the remote miners' rows is gone. Their truly primitive backwardness in regard to sanitary arrangements has received a severe shock. We have not sent so many Labour heavy-weights to the county councils wholly in vain. No longer is the sight to be seen of tumble-down dry closets and open middens scattering dirt and disease over the ghastly square of which they are the principal ornament. If the horrible old rows have still to be lived in, at least their inhabitants have now the benefit of water closets and an inside water supply. The aspect of the Scottish coalfield has been completely altered by the erection on the outskirts of each town and village of the neat, rough-cast, two-storeyed houses which form the county council schemes. A vast deal of demolition and additional building remains to be done, but a beginning has been made on the right lines. An end has been put to the smug indifference regarding these matters that prevailed before the war.

Part of the transformation we are referring to has to be ascribed to the influence of the Miners Welfare Institutes. In many a dull mining village existence during the dark winter months has been made actually endurable by the establishment of a bright, well-lit place of recreation and instruction. Baths, libraries, bowls, billiards, lectures, concerts –these are only some of the multifarious activities of the welfare committees.

If the miner's position as to wages, chances, of hours, and employment is anything but rosy, and is inclining him, it may be temporarily, to lend an ear to the voice of the Communist charmer, he will, at any rate, in his candid moments, be quite willing to admit that his outward environment is, in some respects, a very paradise compared to what it was twenty years ago.[38]

This uplifting view was echoed later in Abe Moffat's recall of the culture of the mining communities. Certainly, heavy drinking and heavy gambling were downsides of the miner's behaviour. Pitching-toss schools, where fortunes were bid and lost on the toss of a coin; and obsessive betting on horses or whippets, could be regarded as negative features. However, as Moffat countered, this was only part of a more complex picture:

Miners liked to go to the whippet-racing, just as the big shots and the Royal Family like to go to horse-racing at Ascot. The only difference is that the latter deal in hundreds and thousands of pounds, whereas the miners, like other workers, have to deal in shillings and on occasion, it may be, a few pounds. Of course it was often said because of this that the miners were heavy gamblers. This was quite the opposite to the truth; no matter what mining village your visited, you would find it was the same small group of miners who were playing pitching toss every week and the same applied to the whippet-racing. The vast majority of miners never played at pitching toss or attended whippet-racing.[39]

Instead, he pointed to the sporting prowess of mining villages and the hosts of footballers and athletes they produced; the popularity of quoit-throwing and of bowling; the pipe bands, brass and silver bands, most of them maintained solely by the mining community; the large number of Burns clubs and community-drama societies.

Political struggle to achieve a better society was Moffat's primary agenda and, while huge industrial defeats were inflicted on the miners, a political transformation was indeed evident during the 1920s. The young Jennie Lee, a miner's daughter from Fife, was elected in North Lanarkshire among a crop of Labour MPs in 1929. However, challenging from the extreme left, the only Communist victory in a parliamentary election was in 1935, in the predominantly mining constituency of West Fife. Willie Gallacher was elected by miners who had rejected the right-wing labourism of Willie Adamson, the long-serving mining union official and former MP. Gallacher was to hold this constituency until 1951, with crucial support from the mining community. During the inter-war period, the mining communities in this constituency formed an exceptional base of political and industrial militancy. Lumphinnans especially, but also Glencraig and Bowhill, had earned reputations as 'Little Moscows' for their record of resistance to the coal companies; their socialist policies on education, housing and welfare; and left-wing, including Communist, representation in local councils and in other areas of community activity. Certainly, the industrial politics of mining unionism in Fife during the 1920s and early 1930s was rather complex, but militant and left-inclined trade unionism was in the ascendant for most of the time. The left-wing Reform Movement had forced a split from the mainstream, moderate Fife and Clackmannan Miners' Union in 1922. It had provided the cohesive fighting force during the 1926 lockout, before rejoining the established union in 1927. Left-wing victories in the union elections in 1927 prompted a right-wing breakaway, led by Adamson. Then, from 1929 until 1935, the Communist-inspired United Mineworkers of

Plate 69.

Gala float at Burghlee, Midlothian, 1917, displaying replica of colliery winding gear. The gala tradition remained a strong feature in mining communities throughout the twentieth century.

(*Collection of the Scottish Mining Museum Trust*)

Scotland was the main trade union presence among Fife mineworkers. Most branches of this militant organisation were in Fife, while several branches also took a hold in Lanarkshire, and a few others in Ayrshire pit villages around Kilmarnock. In Fife, it formed the backbone of campaign support for the Communist candidacy of Willie Gallacher in the 1929 and 1935 General Elections. In those mining localities where it had some influence, this 'red union' provided the principal activists in the National Unemployed Workers' Movement, conducting direct action on several burning issues. For instance, it organised political demonstrations and physical support, at street level, for miners who were sacked and evicted, or who had fallen behind on rent payments due to poverty and lack of work. It represented unemployed claimants before employment tribunals and public assistance officials; it led 'hunger marches' against the Means Test, and advocated the right to paid work at decent wages.[40]

Mining districts like Blantyre and Shotts in the west central coalfield also exhibited notable levels of political and industrial militancy, although they were not free from Catholic and Protestant sectarian rivalry, which continued to sustain political and cultural divisions in that part of the country. Orange elements among miners and contractors in North Lanarkshire had formed a breakaway organisation in the early 1920s to counter a perceived threat of Communist and left-leaning Catholic domination in the Lanarkshire Miners' Union, and their influence had contributed to the return to work before the lockout ended in 1926.[41] However, at other times, from the 1920s onwards, it is not at all clear that workplace solidarity at the coalface was weakened by such divisions and tensions. Instead, the evidence appears to confirm that such differences were dissolved among miners when at work underground. Here, their essential behaviour was conditioned, as ever, by the collective need to observe safety and to look out for each other, and by a common interest in facing up to grievances and injustices.

During the 1930s, the Scottish coalfield was the most strike-prone in Britain, containing a hard core of militant pits, situated mostly in Lanarkshire and Fife, and others in Stirling and West Lothian. Clackmannan, Midlothian and East Lothian were the exceptions, as was Ayrshire, apart from the Kames Pit, Muirkirk. The location and prevalence of militancy in different parts of the Scottish coalfield corresponded quite closely to the strength and capacity of worker leadership and example to resist the policies and practices of intransigent employers, contractors and officials. In most cases, this leadership and example were provided by militants who were in and around the Communist Party. Disputes and worker resistance were also most prevalent in the larger mechanised collieries where several hundred or more workers were employed. Conversely, areas with a proliferation of small pits tended to be management dominated, or had a better history of industrial relations. The Shotts and Blantyre–Hamilton districts claimed the bulk of strikes and disputes within Scotland, and the trouble-torn Bothwell Castle Colliery had the worst record of conflict throughout the decade. Blantyre Colliery held the British strike record between 1936 and 1940, with forty-seven recorded disputes.[42]

A state of guerrilla warfare existed in the most troubled pits, over safety, discipline and dismissal, excessive working time, contractors, speed-up and piece-work rates. None of the causes of strife appear to be directly attributable to the increasing introduction of mechanisation as such, but rather to its effects, particularly to the intensification of work, and constant management attempts to reduce tonnage rates. With managers refusing to recognise trade unions in collieries, the miners had formed unofficial pit committees with shop stewards as leaders. Many of the disputes were short-lived, taking the form of wildcat and lightning strikes, although the numbers of men involved were often high in the larger mechanised pits. Concerted attempts were made, with some success, to eliminate contractors and substitute a fair system of pooling wages in a work team, especially when more favourable conditions permitted after recovery of trade union membership from the mid 1930s. At Polmaise, the colliers were on strike for ten months in 1938 (the longest-running miners' strike in Scotland before 1984–5) in a successful battle to end employment of contractors at the coalface.[43]

This turbulent record of pit militancy continued into wartime (1939–45) as grievances remained unresolved and autocratic managers, their authority boosted by state restrictions on rights of labour in the mines, bullied and cajoled. With the loyalty of rank and file miners to the war effort being unfairly questioned, a rash of strikes and disputes flared up in many Scottish pits, mostly in those pits and areas with a recent tradition of resistance. However, although the several hundred Scottish miners who were members of the Communist Party were influential in leading and supporting militant resistance to grievances in the pits, after Russian entry into the war in June 1941, the party line was changed. Strikes and unofficial action were opposed as means of everyday struggle on wages and conditions, and patriotic duty was declared in favour of all-out production to assist the new wartime ally in the fight against Nazism on the Eastern Front. The bulk of stoppages and strikes in Scottish pits in the final years of the war occurred without official Communist Party support or sanction. Industrial action included huge sympathy strikes of over 10,000 miners in 1943 in support of miners at

Cardowan (Stepps) and Greenrigg (Whitburn) who had been prosecuted for illegal strike action under Order 1305, and who had been imprisoned for refusing to pay the penalty of a fine. This remarkable, and almost forgotten, show of mass defiance in favour of the basic rights of labour compelled the Government to release the Cardowan miners from Barlinnie Prison. State prosecution of miners under Order 1305 was also effectively dropped for the duration of wartime, and was removed from the statute book in 1951 after strong agitation by miners and other trade unions.[44] Meanwhile, the election of a majority Labour Government in 1945 ensured that the inadequate, dual control of the coal-mining industry by the state and the private companies during wartime would end, and mineworkers could reasonably expect improved conditions and prospects under public ownership. The promise and the reality of that experience are examined in the next and final chapter.

Plate 70.

Miners working in 16-inch wet seam, Canderrigg colliery, Lanarkshire, early 1950s.

(*Collection of the Scottish Mining Museum Trust*)

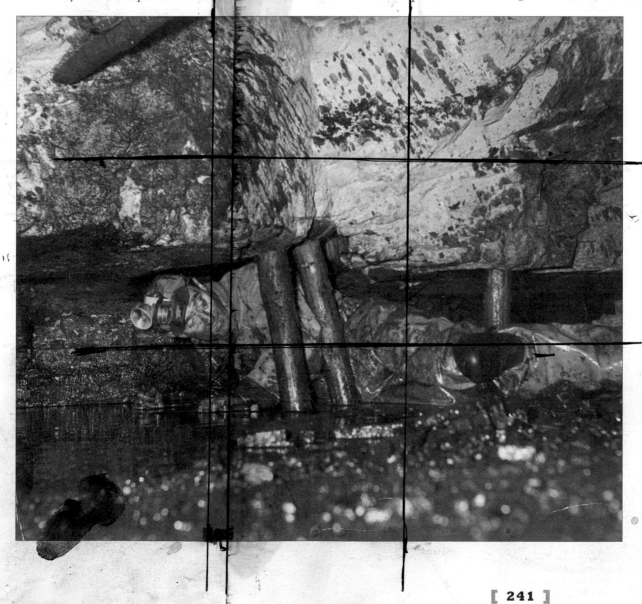

[7] The Final Reckoning: Closures and Conflict, 1947–2002

The Mining Sector: A Nationalised Industry

In January 1947, the Scottish Division of the National Coal Board inherited a fragmented, largely inefficient and rundown coal-mining industry in the throes of long-term decline, although some of the larger companies operated successful collieries and were investing in modern working methods and plant. The new body immediately assumed responsibility for 275 collieries operated formerly by no fewer that 120 separate companies. This legacy included 79 small mines that worked limited areas of shallow and outcrop coal; but the Board decided that this sector could not justify public investment, preferring to leave them to the private owners under a special operating licence. Therefore, discounting the licensed mines, at the outset, the Scottish Division of the NCB directly managed 196 collieries of various conditions and sizes.[1]

Close scrutiny of this core resource confirmed that more than half of the pits, mainly in the Lanarkshire and central coalfield, were found to be already nearing exhaustion of the workable coal measures and/or were old, severely dilapidated and not viable prospects. In this category, over 100 pits were scheduled for phased rationalisation and closure by or before 1965, including an immediate crop. To offset the shortfall caused by heavy output loss from imminent closures, and to begin to raise the overall productive capacity of the Scottish coalfields from 23 million tons to a long-term target of 30 million tons a year, the Board launched an urgent and ambitious expansion programme, as indicated in the first *Plan for Coal* (1950). Accordingly, throughout the 1950s, a massive three-part programme was pursued simultaneously to revitalise and modernise the productive base of Scottish coal. The first undertaking involved the development of around sixty short-life surface drift mines, using the latest technology in high-speed haulage and coal cutters, intended to achieve a quick, reliable source of coal output. The second, and largest, part of the programme consisted of the reconstruction and modernisation of nearly forty viable, long-life pits, mainly in the eastern coalfields. All were extensively re-equipped with additional winding capacity, wider roads to accommodate locomotives and mine cars, and deeper shafts to reach potentially valuable seams at lower depths. This vital component of the programme was intended to produce the bulk of Scottish coal output within a few years. The third component was a long-term project that involved sinking and developing up to nine new, modern super pits, each employing thousands of workers, to pioneer extraction of the mineral wealth in deep-lying seams at 2,500–3,000 feet.

Scotland's Coal Plan (1955) recorded the progress of the reconstruction programme and made optimistic forecasts. However, most of the Board's plans and predictions for this first and only expansionist phase were soon to be dashed, and with them would go the hopes of security of employment and rising prosperity among the majority of Scottish mineworkers. The unravelling of the Board's plans from the late 1950s can be attributed mainly to the effects of various external forces, although some of its troubles were generated internally. Those problem areas included Scotland's permanent geological disadvantage, which hampered profitable working at deep levels; the size and structure of most Scottish pits and the Board's conduct in managing a very diverse coalfield.

Changes in the fuel policy of the Tory Governments of the late 1950s and early 1960s encouraged increased price competition from other, cheaper sources of energy, including a remarkable shift towards the use of gas, oil and electricity in industry and transport. This fundamental change, combined with a minor industrial recession in 1957–8, had a devastating effect on demand for coal. It cut back the vast investment programme for the future of the Scottish mining industry – then estimated at more than £200 million in 1958 – and the largest modernisation programme of all the NCB divisions in Britain. The dramatic shift in the balance of energy consumption and demand away from dependence on coal towards oil and natural gas in the 1960s and 1970s (and later, nuclear power capacity) ended the dominant and historic role that coal had played since the Industrial Revolution in Scotland and Britain. From 1959, NCB policy was adjusted to match output to current demand, and thereafter its revised plans reflected the new reality. During the 1960s in Scotland, this policy change put the expansion and renewal programme into reverse, and accelerated the closure of pits, ostensibly on economic grounds, as being surplus to requirements. These moves were accompanied by furious recrimination within vulnerable mining communities well used to change and dislocation, but who were now facing the premature, unexpected and irreversible loss of thousands of more jobs and livelihoods on top of the experience of enduring the earlier planned closures.

Moreover, compared to England and Wales, the small average size of Scottish mines was an important disadvantage in keeping down productivity and profitability. In 1947, only half of Scottish miners (the overall British figure was 80 per cent) were working in collieries employing 500 or more wage earners. Without the same benefit of 'economies of scale', the average output of Scottish collieries in 1950 was the lowest for any NCB division, despite maintaining the national pace of mechanisation in cutting and haulage. It was also found that more mechanisation, particularly at the coalface, instead of improving productivity, had sometimes actually increased the operational difficulties of working seams in adverse mining conditions, and added to the costs of production. Thus, within the NCB internal market framework, the cost of producing the 'Scottish ton' of staple coal was higher than the British average and was less competitive.[2] In those circumstances, the utmost effort from Scottish mineworkers could not have fundamentally altered that market disadvantage.

While the Scottish Division could not be blamed for this deficiency, stretched and in a haste to make quick progress, it misjudged the contribution of its massive reconstruction and new build programmes towards meeting overall production targets during the 1950s and early 1960s. The modernisation programme was unexpectedly slow and expensive, difficult development work often encountered additional hitches and delays and too high a proportion of the labour costs was swallowed up in payment of contractors and development workers who were not directly involved in coal production. Mining ventures in Scotland, and especially deep-mining projects, always carried higher financial risks due to their greater share of uncertainties arising from peculiar geological conditions. Yet, given the huge, unprecedented costs of developing the larger pits, failures such as the Rothes Colliery were bound to be spectacular and demoralising. Moreover, the scale of investment in all but a few of the long-life deep pits was not to prove justified. The Board was partly to blame, for expensive technical errors, and exaggerated expectations of levels of coal output from key pits over the years.

However, taking into account all the above considerations, it is an inescapable fact that, while there were profitable collieries and units, the mining industry in Scotland as a whole operated at a financial loss from 1950 to at least the early 1980s. Whether, and in what shape, it could have been or become a profitable entity under nationalisation has to remain a matter for conjecture, as also is the contention that it could have survived as a large and viable entity under privatisation. In reality, the most decisive influence dictating the spiral of decline in the coal-mining industry, especially from the 1970s, was to be the impact of Tory government policies on fuel, competition and pricing, and hostility towards coal as a nationalised industry. Latterly, politically motivated attacks aimed at breaking the trade union power of the mineworkers would be equally influential. Then, after the historic defeat of the miners in 1984–5, privatising the industry and implementing the final swathe of closures that would complete the destruction of deep coal-mining in Scotland was to seem an almost inevitable process.

Although the reconstruction and new build programme of the first decade of nationalisation were substantial, they were only partly successful in the longer term, as cuts and closures took their toll. Yet, there were some success stories at individual colliery and area level, and one must include the programme of short-life drift mining, producing expected levels of coal, mostly at relatively low cost. At area level, the Midlothian programme, centred on Bilston Glen and Monktonhall, and the Ayrshire programme, including Killoch, Barony and Littlemill, had the best outcomes over the years.

Of the seven entirely new super-pits, four were good performers, as measured by coming anywhere near their planned potential as long-life coal producers. Bilston Glen tops this short list, followed in order of performance by Killoch, Monktonhall and Seafield (Fife). Among the large, reconstructed collieries, several could be classed as long-term, worthwhile projects, repaying much of the initial capital investment. Kingshill 3, Barony, Cardowan, Bedlay and Polkemmet are prominent in this category.

However, many of the large, new and modernised collieries were hit by early closure in the 1960s and 1970s, before they could achieve their economic potential. Michael Colliery had great potential but, after years of preparation and just as it was ready to go into full capacity to exploit rich seams, its active life was curtailed in 1967 by a disastrous underground fire. Much was also expected from Valleyfield and Kinneil (linked under the Forth) and much was achieved before they encountered geological difficulties at deep levels, resulting in reduced working, and then closure. After nationalisation, the Fife coalfield had been designated as the principal growth area, aimed primarily at expanding exploitation of deep coal reserves in coastal and undersea sections of the Firth of Forth. It was expected to produce the best results, and was allocated the largest share of capital investment for the modernisation programme. Yet, the outcome in the Fife coalfield was assessed 'the most disappointing and least successful' of all the programmes in the planned growth areas within Scotland. Another analyst gave the harsher verdict of 'abject failure'.[3]

Much of the criticism was directed at the Board's worst failure and gigantic white elephant – the Rothes Colliery. This ultra-modern super pit, begun in 1946, opened for limited production in 1956, operational for less than five years and closed and abandoned in 1962, was the most expensive project and a huge loss of potential, regular employment in Fife. The new town of Glenrothes was conceived and started to anticipate the full development and expected long life of the prize colliery. Rothes Colliery was a monument to the gross incompetence of the Board and to the heroism of the men who, in atrocious conditions, built and sank the shafts; drove and constructed the levels; installed haulage, machinery and transport; prepared faces; did emergency maintenance and repair work and produced coal from an upper seam. Sunk on the wrong site, it succumbed to intractable problems, particularly flooding and geological faults.

The positive features of the Clackmannan–Stirlingshire growth area were also blighted by the early failure of two of the projected new super pits in Scotland. Airth was a non-starter, while Glenochil was misconceived and mishandled by the Board.

The reputation of this programme was saved later by the performance of several pits, notably Bogside, Manor Powis, Polmaise and Dollar, in a successful switch to production of low-grade coal to fire the electricity power stations on the Forth. When the large Longannet power station came on stream in the late 1960s, an ultra-modern colliery complex, eventually including Castlebridge and Solsgirth, was created to supply the new venture and the collieries proved their worth.

Before the start of the decisive downturn in 1958, the Scottish coalfield employed 80,000 workers in over 160 active pits. Then, into the 1960s, in all areas, came the greatest massacre of pits, production, jobs and mining communities and in its wake, a trail of dereliction. By 1967, fewer than fifty pits remained, and the workforce had been halved, with the loss of 40,000

Plate 71.

Right. Job losses and output decline in the Scottish Coalfield, 1956–76. (Scottish Miner, *April 1976*)

Plate 72.

Opposite. Pit closures in Scotland 1956–1976. (Scottish Miner, *April 1976*)

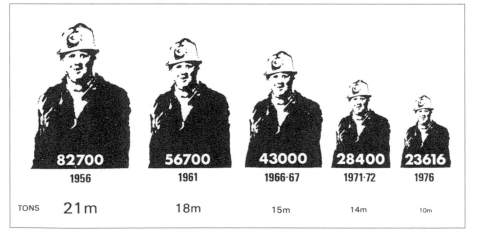

82700	56700	43000	28400	23616
1956	1961	1966-67	1971-72	1976
TONS 21m	18m	15m	14m	10m

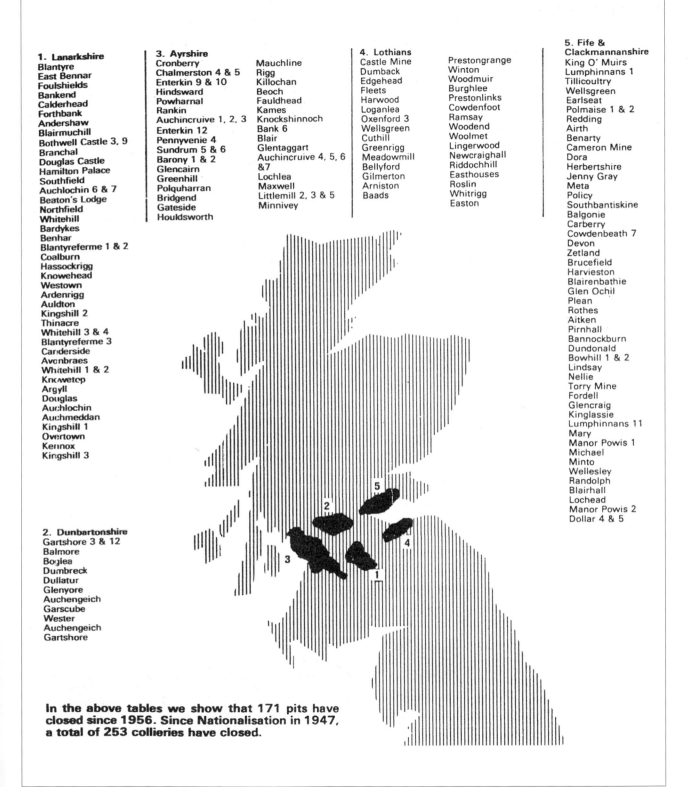

1. Lanarkshire
Blantyre
East Bennar
Foulshields
Bankend
Calderhead
Forthbank
Andershaw
Blairmuchill
Bothwell Castle 3, 9
Branchal
Douglas Castle
Hamilton Palace
Southfield
Auchlochin 6 & 7
Beaton's Lodge
Northfield
Whitehill
Bardykes
Benhar
Blantyreferme 1 & 2
Coalburn
Hassockrigg
Knowehead
Westown
Ardenrigg
Auldton
Kingshill 2
Thinacre
Whitehill 3 & 4
Blantyreferme 3
Canderside
Avonbraes
Whitehill 1 & 2
Knowetop
Argyll
Douglas
Auchlochin
Auchmeddan
Kingshill 1
Overtown
Kennox
Kingshill 3

2. Dunbartonshire
Gartshore 3 & 12
Balmore
Boglea
Dumbreck
Dullatur
Glenyore
Auchengeich
Garscube
Wester
Auchengeich
Gartshore

3. Ayrshire
Cronberry
Chalmerston 4 & 5
Enterkin 9 & 10
Hindsward
Powharnal
Rankin
Auchincruive 1, 2, 3
Enterkin 12
Pennyvenie 4
Sundrum 5 & 6
Barony 1 & 2
Glencairn
Greenhill
Polquharran
Bridgend
Gateside
Houldsworth

Mauchline
Rigg
Killochan
Beoch
Fauldhead
Kames
Knockshinnoch
Bank 6
Blair
Glentaggart
Auchincruive 4, 5, 6 & 7
Lochlea
Maxwell
Littlemill 2, 3 & 5
Minnivey

4. Lothians
Castle Mine
Dumback
Edgehead
Fleets
Harwood
Loganlea
Oxenford 3
Wellsgreen
Cuthill
Greenrigg
Meadowmill
Bellyford
Gilmerton
Arniston
Baads

Prestongrange
Winton
Woodmuir
Burghlee
Prestonlinks
Cowdenfoot
Ramsay
Woodend
Woolmet
Lingerwood
Newcraighall
Riddochhill
Easthouses
Roslin
Whitrigg
Easton

5. Fife & Clackmannanshire
King O' Muirs
Lumphinnans 1
Tillicoultry
Wellsgreen
Earlseat
Polmaise 1 & 2
Redding
Airth
Benarty
Cameron Mine
Dora
Herbertshire
Jenny Gray
Meta
Policy
Southbantiskine
Balgonie
Carberry
Cowdenbeath 7
Devon
Zetland
Brucefield
Harvieston
Blairenbathie
Glen Ochil
Plean
Rothes
Aitken
Pirnhall
Bannockburn
Dundonald
Bowhill 1 & 2
Lindsay
Nellie
Torry Mine
Fordell
Glencraig
Kinglassie
Lumphinnans 11
Mary
Manor Powis 1
Michael
Minto
Wellesley
Randolph
Blairhall
Lochead
Manor Powis 2
Dollar 4 & 5

In the above tables we show that 171 pits have closed since 1956. Since Nationalisation in 1947, a total of 253 collieries have closed.

jobs. The shale-mining industry was also ended in 1962 with the closure of the last shale oil works in West Lothian. In Fife alone, in that decade, the recent, most promising coalfield was reduced to eleven pits, and from 25,000 to 8,000 workers. The grim figures in the accompanying diagrams show the torrid extent of decline and loss over the two decades 1956–76, by which time the Scottish coalfield was down to only twenty-one active pits and around 23,000 workers. The major coalfields were almost wiped out by the end of the 1970s, and the closure of Cardowan in 1983 signalled the final closure of deep mining in Lanarkshire. On the eve of the great confrontation in 1984–5, a rump of fourteen pits and 14,000 workers awaited their fate. Thereafter, the rump became a remnant, with the closure of Bogside, Polkemmet and Frances by 1985; and Comrie, Killoch, Polmaise, Seafield, Barony and Bilston Glen in the late 1980s. Finally, with the closure of Monktonhall and then Castlebridge, deep coal-mining in Scotland became extinct.

Working Conditions and Standards

> When Vesting Day (Nationalisation day) came in 1947, all the pits passed into state hands. We had all been to meetings and listened to speakers telling us about the new Utopia of nationalisation. We were told here would be a rosy future, wages increased, holidays with pay, improved sickness and injury payments – it all sounded good as we discussed it between ourselves. The more cynical asserted there would be no change. It would be the old team in new jerseys, they said, and they were right. Former coalowners were paid vast sums for obsolete pits and even for pits already closed. This was not the nationalised industry we had hoped for and which our fathers had fought for ... and it is not surprising that disillusionment with the NCB set in so soon.[4]

Nationalisation is an important landmark for historians of the labour movement, but for the miners it was another day down the pit. Some things, however, would change for the better, notably the greater emphasis placed on workplace training and on health and safety issues although, as ever, danger stalked the pits.

The calamity at Auchengeich in September 1959, when forty-seven miners died from gas poisoning after an explosion and fire caused by a faulty ventilation fan belt, was the greatest disaster in coal-mining in Scotland during the twentieth century. Another four disasters occurred in Scottish pits between 1947 and 1959, including three from explosions, when firedamp was ignited by the naked flame of a miner's cap lamp. The first occurred at Burngrange shale mine, West Calder, in 1947, killing fifteen men, and was at once the biggest disaster in shale-mining history as well as the first major explosion in a shale mine. In 1957, seventeen men were killed at Kames colliery, and nine at Lindsay colliery. The remaining pit disaster in this early period of nationalisation was at Knockshinnoch colliery, in 1950, when thirteen men were smothered after the roof caved in. They, and another 116 miners, were engulfed by an inrush of earth and moss caused by working a seam too close to the surface. The tragedy apart, this was also the most successful rescue attempt in the history of British mining, as all but the unfortunate thirteen men were saved over the following two days by the heroic efforts of teams of fellow miners.

In all the highly publicised court cases following the coal-mining tragedies above, the NCB, as employer, was found guilty of neglecting and breaking safety regulations, and compensation damages were awarded to the families of the deceased. No breach of regulations was found against the employer in the Burngrange enquiry, as one of the miners had apparently failed to check for gas after a meal break and, in any case, open flame lamps were still allowed in shale mines. Kames was also regarded as a 'naked light

Plate 73.

Throwing lime to suppress coal dust, Wellesley Colliery, Fife (*Collection of the Scottish Mining Museum Trust*)

pit' where safety lamps were not compulsory at all times. However, at Kames and Lindsay, mine officials were found guilty of several breaches of regulations, notably concerning lax conduct of inspections for gas and failing to take adequate steps to suppress accumulation of dust which was liable to set off secondary explosions.[5]

Other major single incidents involving multiple deaths were not officially classed as disasters, but at Michael colliery, in 1967, nine men were suffocated in an underground fire. This colliery, the largest in Scotland, had a complex layout of roads, and management was indicted for failing to erect clear exit signs to places of safety and for inadequate emergency procedures.[6] Nearby, at Seafield colliery, in 1973, five men were killed at 1,600 feet, three miles out under the Firth of Forth, when props and roof collapsed. Seams in the Seafield super pit were notorious for their steep gradients, which, combined with wet conditions, added to the dangers of heavy machine working and internal transport. Machines often came into contact with the roof, and any cave-ins were liable to career down the slope taking everything in their wake. According to Ian Terris, who worked at Rothes colliery and then at Seafield from the late 1960s, roof and stone falls were a frequent hazard. Although he was a qualified shot-firer, he and colleagues were in constant danger from the unpredictable outcomes of blasting operations using powerful explosives in difficult geological conditions. As he relates, his 'guardian angel' saved him from death, though not serious injury, on at least two such occasions.[7] The Seafield disaster was quick to destroy the record set in 1972 – the first year in the history of Scottish mining without the death of a miner from falling roofs of stone or coal.

Avoidable disasters and loss of life, and the continuing toll of needless serious injuries incurred in pit work were a source of grievance among mineworkers and a setback to the high profile of safety concern within the NCB in Scotland. The statistics of fatal, serious and other injuries remained grim. In 1962, designated 'National Safety Year', 40 men were killed, 190 suffered serious injury and another 21, 000 workers were recorded injured (off work for three or more days) in Scottish pits. Thereafter, a downward trend in fatal accidents dropped into single figures each year by the 1970s.[8] In 1974, for example, 7 deaths resulted from pit accidents, but 53 severe injuries, and over 4,000 recorded injuries otherwise blotted an improved safety record, albeit with an ever-diminishing workforce.

However, the bald statistics of death and injury, especially in the 1960s and 1970s, hide a number of complex issues involving health and safety and other pressures at the workplace, and need further explanation. Moreover, they do not account for the incidence of death and disability arising from recognised occupational disease such as the dust-induced condition of pneumoconiosis. Obviously, in the uneven struggle of man against nature in an unpredictable and inherently dangerous underground working environment, a high incidence of accidents and injuries was inevitable. And where human error intervened, casualties would still arise despite the application of best practice in taking precautions, the influence of the many safety campaigns mounted by the employer and by the unions and strict observance of the host of additional safety regulations in place by the 1950s. Risk-taking at the coalface and the foreman sometimes turning a blind eye also accounted for part of that unacceptable injury total. Miners continued to be blamed, with some justification, for carelessness and taking unnecessary risks and short-cuts. However, equally, under modern law the greater onus was on management to provide as safe a workplace as was practically possible, to train up workers properly, to allow scope to carry out tasks well and to enforce safety regulations.

Yet, the high accident rates in Scottish pits under nationalisation into the 1960s and 1970s have to be considered against the background of an intense productivity drive imposed on an ever-decreasing number of workers. Strict enforcement of safety regulations and standards by colliery foremen was one thing: putting pressure on coalface workers to achieve production targets was another matter, and miners could claim that safe working was being jeopardised too often by this conflicting priority. From the 1960s, the planned rise in profitability and output per worker in the Scottish division was to be carried out by a reduced workforce and by investment in the increased introduction of new mining technology. In particular, this consisted of cutters equipped to sheer a whole face of hard coal and stone from roof to floor and power-loading conveyor belts to transport coal directly to the surface. This new integrated cutting and loading machine plant completely changed the pattern of the shift cycle, the nature of the work done at the coalface and the composition of the workforce. Until then, at faces where coal was undercut by machine and then blasted and taken up by a team of manual strippers and fillers and into hutches or conveyor pans, a three-phase shift cycle was usually operated. The first shift prepared the face for the cutting team on the second shift, and the third shift completed the coal removal and cleaned up. However, with the new power technology, the machine operators and the craft electricians and engineers undertook and supervised the cutting and removal of coal on every shift. Support tasks included skilled preparation of road and face, repositioning and maintenance of the machinery, packing the stone waste and propping the roof after the cutter had done its work. A smaller team was required on each shift, the manual strippers and fillers were displaced, and were made redundant or had to adapt to, or be retrained for, different jobs in the coalface squad or elsewhere in the pit. Also, at the pit head, the last manual coal sorters had been displaced by the early 1960s as mechanised screening plant was introduced.

New ways of working, the effects of increased noise and dust pollution, lack of familiarity with complex and dangerous machinery and inadequate training in its handling caused additional hazards and anxieties. Union representatives and workers such as Bob Smith, at Bogside colliery in the 1960s, voiced their concerns:

> Our jobs were changing dramatically, as new machinery came into use. Coal getting now was totally mechanised. Instead of picks and shovels and saws and mash hammers, we had to learn how to use the latest technology in cutters and conveyors. Not all of it was good, and some of it was downright dangerous, exposing the men to roof falls or bringing them far too close to cutter bars and discs. In safety meetings with the management we discussed all of these things, and tried to find ways of modifying them and making them safer. It seemed that all improvements in coal getting brought new problems. The new machines which cut and automatically loaded coal in a continuous cycle were extremely noisy and produced masses of dust. Ventilation was never good, and the air always hot and humid. The air bags which were supposed to control the ventilation and bring fresh air into us were often torn by debris or falling coal, and did not really do their job well. We were supplied with masks but at first they were crude and inefficient. Later models were better, but without doubt they did hinder a man's breathing, and a lot of men found they could not wear the masks when doing heavy work. And, in spite of all the mechanisation, it was still, as always, very heavy work. Everything about a pit is heavy and built to last, and everything is constantly being moved from place to place.[9]

By the late 1950s, more controls on shot-firing, and more sophisticated readings and treatment of gas presence were among the improvements to reduce the causes and effects of unwanted explosions. However, at individual level, more effective and improved design of work-wear was slow to emerge. It was not until 1956 that free safety helmets and boots were issued. Battery lamps replaced carbide lamps only a few years later. Undoubtedly, the latest technology in machine cutting and conveyance meant that miners in the modernised pits were released from winning coal with pick and shovel in cramped and low working conditions which frequently gave rise to beat knee, elbow and forearm injuries. Yet, in surviving smaller and older pits in Lanarkshire and Ayrshire, where coal cutters were rare, hewers and fillers were still enduring such conditions, frequently in kneeling positions, and often in wet seams. It was little consolation to these workers that proper fitting durable knee-pads were designed and made available only from the 1960s. Among the new safety devices, the greatest boon to the miner was the self-rescuer, a small apparatus attached to the belt. It was a virtual life-saver, allowing time to get out of danger. It enabled the miner to breathe properly in a contaminated atmosphere, and to sound an alarm if trapped. After a life of militancy, Mick McGahey claimed in 1987 that the successful conclusion in 1973 of a long campaign to win compulsory wear of the self-rescuer at all times when underground was the highest point of his achievement as leader of the Scottish miners. In evidence before the Michael disaster enquiry in 1968, he had angrily indicted the Coal Board for having halted recent trials of the developing self-rescuer, which, if ready and issued, could have saved the lives of the trapped men in such an instance.[10]

Plate 74.
Self rescuer.
(*Collection of the Scottish Mining Museum Trust*)

One of the problems confronting the miner in his working environment was the underground presence of rats. In all the years when pit ponies had been stabled down the pit, rats had sought to share the straw and other foodstuffs. Weil's disease, which affects the muscles and can be fatal, is an infection contracted from rat urine, and was known to occur among miners. However, in 1951 and 1952, there were outbreaks in Fife, when over twenty miners caught the disease and nine of them died. This crisis led to general rat extermination campaigns. This measure, removal of ponies from the pits and the closure of drift mines and fairly shallow pits where rats congregated most, greatly reduced the presence of rats and by the 1970s Weil's disease was almost unknown among miners.[11]

Dust suppression by water spraying, water jets fitted to cutting machines, and the introduction of face masks were among the measures intended to reduce levels of exposure to the life-threatening dust particles that caused pneumoconiosis. This was the most serious of the lung diseases induced by long-term inhalation and retention of coal dust. First recognised as an occupational disease in 1943 and eligible for compensation, the number of certified cases grew from around 200 cases a year as a result of the intense mechanisation drive in Scottish pits. Until the mid-1970s, Scotland had nearly double the British average mortality rate from this incurable disease. Dust prevention measures were only partly effective. Many miners were reluctant to use masks they considered useless and, despite glowing reports by Mines Inspectors concerning water spraying, jets tended to clog quickly owing to insufficient pressure, water hoses attached to machines were often too short and many faces had to be cut dry. Moreover, the introduction of water spraying was by no means universal, and water injection often produced a sea of black sludge at the coalface, causing yet another source of discomfort.

Despite considerable research into dust disease, chest X-rays were not introduced by the NCB until 1959 and it was the mid-1960s before the first series of regular examinations were completed in all collieries. Certified pneumoconiosis cases were given medical treatment, offered alternative jobs, redundancy and compensation for loss of earnings, although it often took campaigning efforts by union activists to alert and show suffering miners how to apply for their pension and compensation entitlements. In his memoir, Bob Smith also reminds us that union activists also preached vigilance on health and safety matters among their fellow miners, especially concerning better ventilation and dust suppression to reduce the incidence of exposure to industrial disease, while he made himself unpopular when ordering the men to persevere with face masks.[12] While such efforts and radiological detection of occupational lung disease saved lives, especially those of younger workers, they came too late for many older and middle-aged miners and ex-miners already severely disabled and facing premature death. As pneumoconiosis latterly became less of a problem among mineworkers, increasing emphasis was placed on links between the impact of coal and stone dust on other serious lung conditions such as bronchitis and emphysema, both being finally recognised as occupational diseases in the 1990s. The endemic coughing, spitting and disabling breathlessness of the ex-miner still persists. Former miners in Scotland who had contracted one or more of those recognised dust diseases are among the 50,000 or so claimants in Britain currently filing for damages against the defunct National Coal Board and British Coal.[13]

Since nationalisation, miners had continued to make sacrifices in other ways. Although they had won a seven-hour working day in 1947, until 1958 they worked an extra shift on alternate Saturdays to boost coal production. Two weeks' paid holiday was not won until 1951 (effective from 1953) and the 1972 settlement brought a three-week annual holiday. A jumbled mess of local piecework agreements and grades and a poor basic minimum wage for hourly-paid workers guaranteed a host of unofficial disputes and grievances until a more rational wages system was introduced in the 1960s. For instance, major national agreements were negotiated for face workers on power-loading machinery. However, while face workers on production, mine driving and other development work were among the best paid within the industrial sector before the late 1950s, they had slipped down the earnings table by 1970. Until the 1960s, the mining unions had cooperated with the nationalised industry by holding back higher wage claims in order to win deals to halt closures. From then on, this policy was increasingly rejected as futile and self-defeating while closures continued unabated and jobs were lost forever. The national wages strikes of 1972 and 1974 were conducted in a mood of resurgent militancy, and were won by determined and well-organised picketing amidst considerable public sympathy for the claims of the mineworkers.

Final Spiral of Closures and Conflict

During the long-term decline of mining in Scotland from the 1920s, mineworkers and their families had become used to the disruption of frequent closures, changing jobs, moving company house and migrating short or longer distances. From 1926 to 1947, under private ownership, no fewer than 600 pits were closed down in Scotland, and thousands of mineworkers had left the industry, many to join the emigration queues to Australia, New Zealand or North America. Before 1914 and again in the 1930s, significant numbers of Lanarkshire miners had migrated to Fife, and smaller numbers to the Lothians, where pit work was more plentiful. More often though, prior to 1947, miners affected by closure who somehow still managed to stay in the industry did not always have to move house or away from the immediate locality. Instead, sooner or later, many chanced either to find work with the same company at another pit, or with another owner, within tolerable travelling distance by bicycle, bus or rail.

During the first decade of public ownership, closures and redeployment of miners were less haphazard experiences, as the Scottish division of the NCB assessed its resources to increase overall coal production and maintain profitability. It carried out a coordinated programme across the whole coalfield, including planned and phased closures, prior consultation with the workforce and assisted schemes for transfer and reallocation to the more viable pits and growth areas. A large internal transfer scheme was designed to reallocate miners displaced by planned closures throughout Scotland. Of nearly 7,000 miners redeployed from the 76 closures between 1947 and 1956, over 5,000 were absorbed and retained at other collieries, while just over 1,000 were granted redundancy benefit or premature retirement. Situated in a declining coalfield, Lanarkshire miners were most affected by targeted closure. Before 1958, the bulk of around 3,500 displaced miners in Lanarkshire were redeployed locally to other collieries. However, by 1956, under NCB auspices, another 7,000 miners and their families migrated from Lanarkshire to work and settle elsewhere. This number, destined mainly for Fife, others for the Lothians and Ayrshire, included a minority of several hundred miners who moved as a direct consequence of redundancy; but the majority of migrants were working miners who volunteered to transfer under the assisted scheme. Young married men in the prime of their productive working lives, and seeking new prospects for themselves and their families, responded to the NCB initiative to attract this category of young and skilful coalface worker to augment the additional manpower needs of the designated growth areas. Key incentives were the promise of a modern rented house and security of employment in a long-life colliery.

The NCB in Scotland claimed credit for resourceful, sensitive and successful handling of the voluntary transfer scheme and the labour retention scheme from closing collieries during the first decade of its management. Certainly, the Board was prompt and vigorous in resourcing its labour schemes. Transfer allowances were in full operation by 1948, and consultation with miners and families included free visits to likely collieries and areas. Hostels provided temporary accommodation for transferred miners; there was considerable headway with plans and provision of housing and sites in conjunction with local authorities and the Scottish Special Housing Association; and subsidised transport was issued to home-based miners.

A programme of 13,000 new houses was implemented by 1955 for miners transferred or already settled in the receiving areas. However, despite its achievements in managing change during this period of consolidation, the Board's claims of a smooth transition – 'the migration proceeded without interruption and with complete satisfaction to all concerned', and 'married couples were found to speak highly of the change' – were disputed by miners and by their trade unions.[14] For instance, a study of Shotts miners and their responses to the transfer schemes before 1951 revealed the extent of disillusionment among many who moved to Fife and the Lothians.[15] Although the disgruntled were in a minority, they found that the promised houses and amenities were not ready or not in place. Those in new houses found a difference in the cost of living, having to rent and maintain a modern house on wages which were no higher than they had been used to; and members of families were often unable to get jobs in the new localities. Several returned to the home area to resume their chances of work and housing in a familiar environment with a thriving community life. Such objections were well founded as, in contrast, new enlarged villages like Ballingry, in Fife, built mainly for incoming families of miners, were still little more than big housing schemes in 1955, with few shops and no community amenities.

In any case, the prospects of so many migrant miners already satisfied by the changes were to be dashed during the second decade of NCB management, as the rundown of coal production and widespread closures increasingly threatened mass redundancy. Between 1957 and 1966, a tidal wave of 124 closures directly affected the livelihood of 30,000 miners in Scotland, predominantly in Lanarkshire and in Fife, which had been transformed into a declining area. Between 1962 and 1966, the transfer policy was extended beyond Scotland, and over 3,000 miners moved south to more prosperous mining areas, including 1,400 to Yorkshire, and nearly 1,500 to the Midlands. During the 1960s, other industries absorbed some of the surplus labour from mining; for instance, the Ravenscraig steel complex provided work for many former Lanarkshire mineworkers. To be fair to the Board, it managed to redeploy many thousands of its displaced workers, while many older and unfit workers were pensioned off.

The leadership and local area committees within the National Union of Mineworkers in Scotland had contested closures from the beginning. While they did not object to closure of pits on grounds of proven exhaustion, they disputed other closures, on the basis that increased production of coal was still government policy and in the public interest. They rejected the profit and loss accounting that dictated the fate of a working pit or group of pits as being 'uneconomic'. They also warned against the dire social consequences of clusters of closures in mining areas like Shotts, where little alternative employment existed and neither transfer nor reallocation was a realistic or desired option for an ageing workforce.[17] According to the criteria of the Board in the 1950s, the cost of continuing production in typically small and modest size, old and difficult to work pits, as in the whole Shotts mining district, was uneconomic and they were, therefore, destined for closure. Campaigns to prevent or halt closures were vehement and vociferous but were eventually a lost cause; and, in the Shotts case, its fate as a mining community was sealed within eleven years, all nineteen pits being closed between 1949 and 1960.

Thereafter, into the 1960s and 1970s, pit closures, especially in Lanarkshire, Fife and Ayrshire, brought redundancy instead of redeployment, as job loss turned into mass unemployment, contributing to the de-industrialisation of central Scotland in the final decades of the twentieth century.

Plate 75.

The Scottish pits in 1979.
(*J McCormack*, Polmaise)

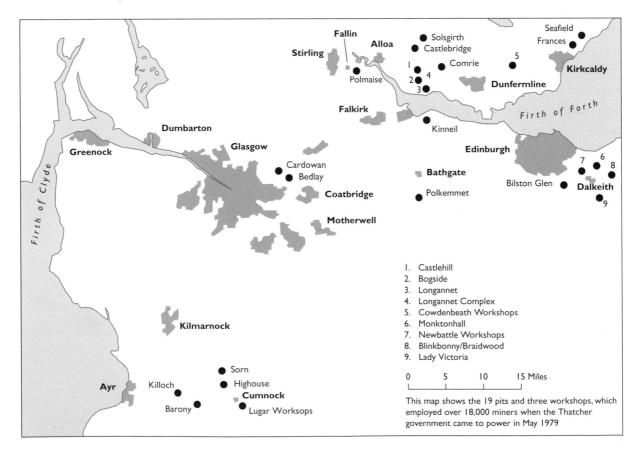

1. Castlehill
2. Bogside
3. Longannet
4. Longannet Complex
5. Cowdenbeath Workshops
6. Monktonhall
7. Newbattle Workshops
8. Blinkbonny/Braidwood
9. Lady Victoria

0   5   10   15 Miles

This map shows the 19 pits and three workshops, which employed over 18,000 miners when the Thatcher government came to power in May 1979

This strike was not about wages and conditions – it was a struggle to defend the very existence of mining communities and it would be fought with bitter intensity to the very end.[18]

We were solid during the strike at Polmaise. There was not a single attempt at scabbing throughout the whole year. No-one even tried to get to their work, so we didn't need a picket line.[19]

The women mair or less started going up to the strike centre to help wi' the meals. As the strike progressed, women being women, they started to make decisions and it wasnae accepted too well tae start wi'. But we organised a women's picket, maybe a month tae six weeks intae the strike. An' we went oot wi' oor banners and oor placards tae the Killoch and actually turned every fireman, oversman, and a' the officials that day and we got a guid reception fae the men when we came back. So I think they appreciated then we had a place.[20]

A comprehensive social history of the miners' strike in Scotland has yet to be written, although many powerful impressions of that experience are already available from those who were directly involved, providing valuable insights into the struggle in various mining localities.[21] The selected statements above exemplify the principled stand taken by the majority of Scotland's 14, 000 miners and their families, as they joined around 100,000 others in the English and Welsh coalfields in the heroic year-long struggle to fight closures and save livelihoods.

The apparent response to the strike call among miners in Scotland was generally one of determined resolve, and remained so until at least early summer in all mining localities. However, as the strike dragged on unexpectedly and into the winter of 1984–5 with no prospect of concessions by a Tory Government visibly intent on smashing the strike and the NUM, it became increasingly difficult to sustain morale. While many mining communities and individual miners and families remained resilient until the bitter end, the brave united front had been breached, as substantial numbers of miners succumbed to the pressure to return to work before the end of the year.

The absence of a national strike ballot was a most controversial and divisive issue, not least within the mining community where opinion was adversely influenced by government and media clamour for a ballot, accompanied by their campaign to discredit the leadership of the NUM and Arthur Scargill, in particular. A fully committed activist from Scotland provided a clear and informed explanation of the principles and the dilemma:

> The absence of a ballot ... haunted the NUM throughout that year. It provided an excuse for some and a rod for others and the miners were spared neither for those twelve months. Within the rule book of the NUM, specifically Rule 41, the strike was perfectly constitutional. Two areas, Scotland and Yorkshire, called for action under their own rules and the union's National Executive declared it official. In time-honoured, and democratic tradition, picketing began and the strike spread according to the principle of the domino effect.
>
> Yet, tactically, there should have been a ballot. It was a catastrophic mistake not to call one, particularly after a special conference in April 1984 had reduced the majority required from 55 per cent to 50 per cent plus one. No one doubts a majority would have voted to continue the strike; we were already dug in by that time.
>
> In the strike centres we subscribed wholeheartedly to the idea that a man working in a coalfield safe from the threat of unemployment should not be allowed to vote another man out of his job. But the absence of that ballot did not, for us, remove one shred of legitimacy from our action. It was right to be on strike and the main task was to strengthen the forces already marshalled on the ground. By the end of April the question had become academic to the vast majority of those miners on strike, who constituted the vast majority in the coalfield, yet it was an opportunity lost and it cost the union dearly, financially and morally, for the rest of the year.[22]

The strike action had started at Polmaise in March 1984, as miners voted to defend their pit against expected closure. The Board had announced closure in December, claiming that there was no market for the coal, despite spending £15 million on the development of new seams ready to resume production for power stations. For the NUM leadership, targeted closure of a developing and economically viable pit like Polmaise, and the threat to others, were ample proof that no pits were safe in Scotland. At this late stage, there appeared little choice but to mobilise trade union power to defend their industry and repel the menace of hard-line Tory policies.

The strike was conducted against tremendous odds, on various fronts. Firstly, the Tory Government, led by Margaret Thatcher, re-elected with a large majority in 1983, had a battery of anti-union laws to use against the striking miners and to deter any other trade unions from taking supportive strike action. Workplace picketing was limited to six union members only, and picketing of other workplaces such as steelworks, depots or docks was outlawed. Thus, to be effective, direct action by miners at pit gates and elsewhere was bound to come into conflict with the full force of the law. Secondly, the decision by NACODS (the trade union of pit deputies, colliery officials and shot-firers) not to join the strike was a definite setback, although the NUM never refused to do safety cover. Thirdly, while activists and local branches within the labour and trade union movement in Scotland were generous with cash, provisions and other material and moral support, the contrasting failure of the Labour leadership and trade union chiefs to back the miners with industrial action and challenge the anti-union laws was a more decisive setback. This refusal to take action was widely condemned by striking miners and support groups as a betrayal, allowing the Tory Government to take revenge on the miners for the humiliating defeats of 1972 and 1974.

Lastly, striking miners and their families had to make ends meet on the home front. Under new, more stringent social security clauses, strikers' families were deemed to be receiving strike pay, and benefit was cut accordingly by £15 (later £16) a week, less than half the amount normally paid in state benefits to an out-of-work family with two children. With no earnings coming in for the best part of a year, hardship was inevitable in many families, not forgetting young single men who had no entitlement to welfare assistance and had to look to family and friends for support.

> Coping with the strike drew virtually every miner's family into contact with the DHSS and the Social Services. For some this led to the most distressing and humiliating moments of the strike. One woman who applied for a single payment to buy maternity clothes was told by the DHSS that she could use a pin to hold her skirt or trousers. Later on in the dispute other women were more fortunate and received money to buy maternity clothes. The pattern for other claims was similar. Sometimes a claim, for example, for replacing a cooker was granted, sometimes not.[23]

Labour-controlled local authorities in Scotland were generally supportive. For example, Stirling and Midlothian district councils gave rent rebates and free leisure and recreation facilities to striking miners and their families, and regional councils gave emergency loans. Fife council paid out over £1 million

Plate 76.
Mick McGahey,
NUM leader.

in loans to over 2,000 mining families over the year of the dispute. The amounts from Strathclyde Region's social work department paid to mining families in Mauchline, Ayrshire, 'ranged from £150–£1,500 according to circumstances and the number of dependants. These loans were not made in a lump sum but were made up of several payments. The applications for relief loans assumed spending of £20.80p per week on food, heating and clothing for three.'[24]

At local level, the strike centres were the organisational and resource hubs, responsible for co-ordinating and carrying out many support activities. They arranged picketing rotas and transport, meetings and publicity; provided advice and assistance on welfare benefits, hire purchase, rent and mortgage payments; organised legal advice in court cases arising from debt arrears and police charges for alleged picketing offences; acted as a sorting and distribution centre for hardship funds, donations of cash, food and presents; and organised outings for the children and old folk. Most centres also provided the vital service of communal feeding, in an updated version of the soup kitchen, for all striking miners and families who wanted to take advantage of this service.

While the strike was organised by the NUM activists at area and local level, the involvement of women was vital at home, in the various strike centres and in support groups. It is widely recognised that without the

Plate 77.
Confrontation, Bilston Glen Colliery, 1984–85.
(*Collection of the Scottish Mining Museum Trust*)

women's support, the strike could not have been maintained for so long. We have shown earlier how this measure of support was not a new feature in mining communities, but on this final occasion it took on a completely new dimension. There were perhaps as many as ninety women's support groups in all the Scottish coalfield areas during that strike year, working mostly in close conjunction with the NUM. For example, the Lothian Women's Committee coordinated twenty-eight local support groups set up by wives of miners from Bilston Glen, Monktonhall and the Newtongrange workshops. From Fife, a miner's wife and activist explained:

It was the women who organised themselves, but in getting off the ground they were given a great deal of support by the local NUM. They helped us set up our organisation and, whenever we needed it, provided transport to meet women from the ten centres coming under the Dysart umbrella. We began to realise over a short period what a tremendous impact the women's groups were having, not just in cooking meals and raising funds, but as a moral booster for our men themselves. Soup kitchens and bingo nights soon became insufficient for many women. The men certainly realised the importance of involving the women, at least the local committee did. We started visiting factories and union meetings; we were invited to speak at meetings, a skill many of us acquired quite speedily. Wherever the men went to speak the women went too.[25]

The same commitment was shown elsewhere, although

it was frightening for housewives that had never been used tae public speaking tae get up there and dae their bit, but they did it. We even had lassies getting sent tae Belfast, Queen's College, tae address a rally. I was at Belfast masel, but it was an open air rally. It was on the back o' a trailer and ma knees were shaking. A' thae things were the good points, broadening the lassies' horizons and making us mair politically aware.[26]

Although many strikers were opposed to or anxious about their wives being involved in pickcting, on a limited and agreed extent, women joined the picket lines, and some were arrested for minor offences along with the men. A contingent of women, including members of the Lothian groups, took part in the mass picket at the Ravenscraig steel works. They found it a frightening and politically enlightening experience being confronted by greater numbers of police (and, as they claimed, soldiers in police uniform) and pushed up against the fences.[27]

In the Stirling area, mining villages like Fallin and Cowie were strongholds of collective solidarity and resilience throughout the year-long strike. In the Fife coalfield, the tradition of class-conscious militancy and resistance prevailed, as only a few men ever defied the strike and went in to work. However, at the four large pits in Lothian and Ayrshire, after being initially solid, a minority had crossed picket lines, and Bilston Glen and Monktonhall were the scenes of most of the incidents and arrests in Scotland as mass pickets were confronted by systematic police operations. Killoch and Barony were also targeted for extra picketing as the strike wore on and as resolve began to wilt in some of the more isolated Ayrshire centres.

According to several accounts, the NCB had many tactics for exerting pressure and inducing men back to work. In Ayrshire, some went back to work 'lured by the promise of Christmas bonuses, holiday pay, and tax-free wages until the end of the tax year'.[28]

Improved redundancy offers towards the end of the strike undoubtedly were influential in persuading older men to return, and there were also rumours and fears that being strike-bound for a year might endanger entitlement to benefits or renewal of work contracts. Jackie Aitchison, NUM branch secretary at Bilston Glen, was convinced that the Board did all it could to break the strike there:

> We were obviously the largest colliery, the largest unit – if you break that, then there was every chance that the whole of the Scottish strike would fold as well. But certainly the pressures at Bilston Glen were at times horrendous. Every new angle that British Coal could think of to break the strike was tried at Bilston Glen first: personal contact by senior management, chapping men's doors late at night asking them to come back; inducing them with various monetary promises; attempting to discredit union officials at every possible turn; arrests that took place on trumped-up charges that were later dismissed from court.[29]

There were 1,350 police arrests in Scotland for alleged picket line incidents, despite the fact that the strike was solid for so long in most places, and mass picketing was not used or not often required. The charges were for minor breach of the peace, unlawful assembly, obstruction of police or highway and assault. Many charges were thrown out, only 470 cases came before the courts and only one prison sentence was imposed, although heavy fines were a regular feature. David Hamilton, leading NUM activist at Monktonhall and chairman of the central strike committee was remanded for three months on a serious assault charge. After a two-day trial, it took a jury only twenty minutes to find him not guilty. Evidently, accusations against the miners about the severity of picket line violence were distorted and exaggerated for political reasons, while dubious police behaviour was rarely called to account.

The biggest confrontations with police arose in May at Ravenscraig, and at the Hunterston terminal, where imported coal stocks were unloaded, destined for the steelworks. Here, on 8 May, around 1,500 pickets faced over 1,000 police and, after a police cavalry charge, 65 arrests were made in one of the most critical and criticised incidents of the conflict.

> The miners lined up and the police horses simply went right into them. I was surprised no one was actually killed. My brother James was among the 65 miners lifted at Hunterston. He was flung into jail and kept there for 10 hours, and ended up being fined £150 for breach of the peace. His 'crime' was that he was standing there when the horses charged and was knocked down.[30]

The same eye-witness called this incident a 'miniature Orgreave', foreshadowing the more infamous incident at Orgreave, Yorkshire, in June, where it has been subsequently proved that the so-called riot was engineered by the police.[31]

In another major incident, mass pickets from Fife and the Lothians converged at Lochgelly to stop the NCB from shifting opencast coal from the nearby site to power stations in England. 'Whilst the picket was relatively peaceful, nevertheless the police took a heavy-handed approach, arresting almost anyone who moved, so that in two days 133 pickets had been charged with obstruction. This was the greatest number of arrests at any one picket in Britain.' However, small victories were possible, as shown in this continuing episode. When the Yuill and Dodds lorries attempted to reach Lochgelly by going through Ballingry,

> they were met with great hostility by pensioners, housewives and youngsters who poured abuse on them at every point along the road. The final straw for the lorry drivers came when well over 100 youngsters, between 12 and 14 years, instead of going into school, marched out on to the road down through the village to Lochore Miners' Institute, singing and shouting and blocking the way of the worried lorry drivers. They never re-appeared.[32]

Although the police actions at confrontations such as Hunterston and Orgreave were controversial, they turned the tide against the tactic of the mass picket and mining communities then sensed they could not hope to win if left to battle alone. It was like 1926 all over again, as the NCB in Scotland sacked and subsequently refused to reinstate over 200 miners who had been criminalised for picketing offences and other alleged troublemaking, including many who had been exonerated of any charges. The highest number of victimised men from any one colliery in Britain was the forty-six at Monktonhall, including all the activists, the elected NUM branch committee and strike committee members. By such action, the NCB had an obvious interest in destroying the influence of the NUM in Scotland and imposing uncompromising managerial rule over the remaining workforce. The campaign to win justice for the sacked miners, led by the rump of the NUM and the remnants of women's support-group activity, was also a matter of public concern for several years. By 1988, some justice had been achieved for most of the victimised men although, for the others, the issue was to remain an unresolved personal and family loss and, for the labour movement, a scar on its conscience.[33]

Postscript

The miners' strike was the last ditch struggle of the older working class for survival in a capitalist state that had long ceased to regard its industrial workers as indispensable. In the Scotland of the 1980s and 1990s, the deep coal miner was soon to become an industrial relic, alongside the shipbuilder, the steelmaker and allied engineering trades. Final closures condemned most of the remaining pit villages to become unemployment blackspots and to a pervasive sense of drift and emptiness as redundant and defeated miners and their families tried to pick up and restore their lives. The ever-decreasing demand for coal into the foreseeable future of the twenty-first century is met by foreign imports and a privately owned opencast sector that employs large excavating machines and few workers. Revival of deep coal mining in Britain is not on the agenda of big business or any of the political parties, although the arguments rage on about its past and its prospects. Now, it is aggrieved residents on or near former mining sites in central Scotland who are the reluctant warriors, as they protest with their allies in the environmental movement to stop the local spread of opencast working, its ravages and pollution, and toxic landfill waste dumping.

Notes

Chapter 1

As a starting point, the early history of mining in Scotland is summarised in J. Mckechnie and M. Macgregor (eds), *A Short History of the Scottish Coal-Mining Industry* (Edinburgh, 1958). J. U. Nef, *The Rise of the British Coal Industry*, 2 vols (London, 1932) provides considerable detail, but has to be used with caution. Its coverage of the Scottish coalfield is superseded by more recent scholarly treatment in J. Hatcher, *The History of the British Coal Industry, vol. 1. before 1700* (Oxford, 1993) and C. A. Whatley, 'New Light on Nef's Numbers: Coal Mining and the First Phase of Scottish Industrialisation, *c.* 1700–1830', in A. J. G. Cummings and T. M. Devine (eds), *Industry, Business and Society in Scotland Since 1700* (Edinburgh, 1994). For the lead industry, T. C. Smout, 'Lead Mining in Scotland 1650–1850', in P. L. Payne (ed.), *Studies in Scottish Business History* (London, 1967), is essential reading.

Smout's *A History of the Scottish People 1560–1830* (London, 1969), pp. 178–83, is an eloquent account of the foundations of colliery serfdom. B. F. Duckham, 'Serfdom in Eighteenth Century Scotland', in *History*, 54 (1969), pp. 178–81 is most useful in this context.

For new critical insights into the status of colliery workers and mining communities in the era of serfdom, in addition to his article cited above, Chris Whatley has contributed several other important articles. As used here, they include: '"The fettering bonds of brotherhood": combination and labour relations in the Scottish coal mining industry *c.* 1690–1775', in *Social History* 12:2 (1987); 'A Caste Apart? Scottish Colliers, Work, Community and Culture in the Era of "Serfdom," *c.* 1606–1799', in *Scottish Labour History Society Journal*, 26 (1991); and 'The Dark Side of the Enlightenment? Sorting out Serfdom', in T. M. Devine and J. R. Young (eds), *Eighteenth Century Scotland: New Perspectives* (East Linton, 1999). His 'Salt, Coal and the Union of 1707; A Revision Article', in *Scottish Historical Review*, 66 (April 1987) and *The Scottish Salt Industry: An Economic and Social History* (Aberdeen, 1987) explain the salt and coal connection, and Chapter 5 of the book reveals the status of the salt workers.

1. A. I. Bowman, 'Culross Colliery. A Sixteenth Century Mine', in *Industrial Archaeology*, 7:4 (1970), pp. 353–72.
2. Extracted from P. Hume Brown (ed.) *Early Travellers In Scotland* (1891, reprinted Edinburgh, 1973) pp. 116–17.
3. Quoted in N. Davidson, *Discovering The Scottish Revolution 1692–1746* (London, 2003), pp. 33–4.
4. Whatley, 'The Dark Side', p. 261.
5. Whatley, 'The Dark Side', p. 262.

Chapter 2

Robert Bald, *A General View of the Coal Trade of Scotland* (1808) is a major primary source for this chapter. Ninth Earl of Dundonald, *Description of the Estate, particularly of the Mineral and Coal Property … at Culross* (1793) has also been directly consulted (National Library of Scotland copy). Two collections of primary source materials have also proved useful: R.H. Campbell, and J.B.A. Dow, *Source Book of Scottish Economic and Social History* (1968); and Elliott, B.J. (ed.), *The Coal Industry: 17th-20th Centuries* (Central Regional Council Education Department,1978), pamphlet No.2 in series of historical sources for schools.

There is a huge secondary literature on Scottish industrial development for this period. For general coverage, including coal mining, see again Smout (1969) pp. 430–40. Whatley, *Scottish Society 1707–1830* (2000) is the best survey; and A. Slaven, *The Development of the West of Scotland: 1750–1960* (1975) remains a very useful regional study.

B.F. Duckham, *A History of the Scottish Coal Industry, vol. 1, 1700–1815* (1970) is a major source for this chapter, as is Flinn, M.W., *The History of the British Coal Industry vol. 2, 1700–1830* (1984). On labour, community and industrial relations in mining, see the articles by Whatley, as cited above for Chapter 1. Two useful case studies of the Clerks and the mining community at Loanhead are: Duckham, B.F., 'Life and Labour in a Scottish Colliery 1698–1755', in *Scottish Historical Review* XLVII, 2 (1968), and Houston, R.A., 'Coal, Class and Culture: Labour Relations in a Scottish Mining Community 1650–1750', in *Social History* 8, (1983). Campbell, A.B., *The Lanarkshire Miners: A Social History of their Trade Unions 1775–1874* (1979) is the indispensable text on this topic, with coverage beyond Lanarkshire for the earlier period. Hassan, J.A., is the leading authority on the Lothian mining community, contributing 'The Landed Estate, Paternalism, and the Coal Industry in Midlothian 1800–1880', in *Scottish Historical Review* 59, (1980).

Douglas, R., 'Coal Mining in Fife in the Second Half of the Eighteenth Century', in Barrow, G.W.S., (ed.), *The Scottish Tradition* (1974); Campbell, R.H., *The Carron Company* (1961); and Whatley, C.A., ' "The Finest Place for a Lasting Colliery". Coal Mining Enterprise in Ayrshire c. 1600–1840', in *Ayrshire Collections*, 14, 2, (1983), are other useful local studies of mining developments.

1. Bald, *General View of the Coal Trade of* Scotland, pp. 87–91.
2. Duckham, (1970) *History of the Scottish Coal Industry*, pp. 29–32.
3. Elliott (ed.), *Coal Industry*, p. 13; extracted from *The Times,* 29 April, 1805.
4. Duckham, *History*, (1970) p. 72.
5. Elliott (ed.), *Coal Industry*, ed pp. 13–14; letter from I. R. Bald, 1816.
6. Campbell and Dow, *Source Book*, pp. 184–5.
7. Flinn, *History of the British Coal Industry*, pp. 359–60.
8. Duckham, *History*, (1970) pp. 267–8.
9. Whatley, 'A Caste Apart?' (1991)
10. Campbell and Dow, *Source Book*, pp. 140–1.
11. Elliott (ed.), *Coal Industry*, pp. 10–11.
12. Campbell and Dow, *Source Book*, pp. 141–2.
13. Campbell and Dow, *Source Book*, pp. 144–5.
14. Duckham, *History*, (1970) p. 310.
15. Dundonald, pp. 68–9.
16. Duckham, *History*, (1970) p. 292.

17. An Address to the Colliers of Ayrshire at the Formation of the Colliers' Association (1824); also discussed in Campbell, A.B., *The Lanarkshire Miners*, pp. 63–4.

Chapter 3

The key primary source is the Children's Employment (Mines) Commission, *Parliamentary Papers* 1842, vols XV and XVI. The reports and evidence submitted by the two sub-commissioners for Scotland are in vol. XVI, which is the Appendix to the First Report, Part 1.

The other major primary source is again Bald, *A General View of the Coal Trade of Scotland* (Edinburgh, 1808), especially in the section containing his enquiry into women bearers.

A. Miller, *Coatbridge, its Rise and Progress* (1964) includes important contemporary comment on the Monklands mining and industrial community.

Secondary works consulted included Flinn, *The History of the British Coal Industry, vol. 2, 1700–1830* (Oxford, 1984), and the next volume in this series, R. Church, *The History of the British Coal Industry, vol. 3, 1830–1913* (Oxford, 1986). A.B. Campbell, *Lanarkshire Miners* (Edinburgh, 1979) and G.M. Wilson, *Alexander McDonald, Leader of the Miners* (Aberdeen, 1982) discuss the work of male colliers and ironstone miners.

J. A. Hassan, *The Development of the Coal Industry in Mid and West Lothian, 1815–1873* (Strathclyde University Ph.D thesis) provides essential material on work and social relations; and P. L. Payne, 'The Govan Collieries 1804–5', in *Business History* 3 (1961), pp. 75–96 is a snapshot of a modern colliery of the time. Two pioneering books, R. P. Arnot, *A History of the Scottish Miners* (London, 1955), and I. Pinchbeck, *Women Workers and the Industrial Revolution 1750–1850* (London, 1930; third edition, 1981) are still useful reference points for the Children's Employment Commission report and its exposure of working conditions. L. King, *Sair, Sair, Wark* (Kelty, 2001) is an informed and committed popular study of women and mining in Scotland. Her Chapters 2–5 cover women and child labour to the 1840s.

1. Bald, *A General View of the Coal Trade of* Scotland, pp. 141–2.
2. 1842 report XVI, p. 436; the first witness interviewed, employed at Sherriffhall Colliery (owner, Sir John Hope).
3. From conclusion of report submitted by R. H. Franks to the Children's Employment Commission, 1842, p. 347.
4. Bald, *A General View*, pp. 43, 74.
5. 1842 Report XVI, pp. 326–7.
6. Miller, *Coatbridge*, p. 187.
7. Campbell, *Lanarkshire Miners*, p. 41.
8. 1842 Report XV, p. 96.
9. 1842 Report XVI, p. 489.
10. 1842 Report XVI, p. 503.
11. 1842 Report XVI, p. 436.
12. 1842 Report XVI, p. 436.
13. Bald, *A General View*, pp. 74–5.
14. Whatley, 'Dark Side', p. 266.
15. Bald, *A General View*, pp. 79–80.
16. Wilson, *Alexander McDonald*, p. 12.

17. 1842 Report XVI, p. 326.
18. 1842 Report XVI, p. 327.
19. 1842 Report XVI, p. 472.
20. Bald, *A General View*, pp. 131–7.
21. 1842 Report XVI p 458.
22. 1842 Report XVI, p. 458.
23. 1842 Report XVI, p. 446.
24. 1842 Report XVI, p. 447.
25. 1842 Report XVI, p. 440.
26. 1842 Report XVI, p. 448.
27. Bald, *A General View*, pp. 140–1.
28. 1842 Report XVI p 460.
29. 1842 Report XVI, p. 460.
30. 1842 Report XVI, p. 458.
31. Dundonald, *Description of the Estate, particularly of the Mineral and Coal Property … at Culross* (1793), p. 55.
32. Bald, *A General View*, pp. 137–9.
33. 1842 Report XVI p 509.
34. 1842 Report XVI, p. 383.
35. 1842 Report XVI, p. 324.
36. 1842 Report XVI, p. 388.
37. 1842 Report XVI, p. 479.
38. 1842 Report XVI, p. 449.
39. 1842 Report XVI, p. 479.
40. 1842 Report XVI, p. 501.
41. 1842 Report XVI, p. 444.
42. 1842 Report XVI, p. 456.
43. 1842 Report XVI, p. 509.
44. 1842 Report XVI, p. 510.
45. 1842 Report XVI, p. 510.
46. 1842 Report XVI, p. 438.
47. 1842 Report XVI, p. 473.
48. 1842 Report XVI, p. 449.
49. 1842 Report XVI, p. 449.
50. 1842 Report XVI, p. 449.
51. 1842 Report XVI, p. 324.
52. 1842 Report XVI, p. 444.
53. 1842 Report XVI, p. 362.

Chapter 4

1. J. Hamilton, *Poems, Essays and Sketches* (Glasgow, 1880), p. 75.
2. A.B. Campbell, *Lanarkshire Miners*, chapter 4; and G.M. Wilson, *Alexander McDonald*, chapters 2 and 3, provide essential coverage.
3. 1842 Report, XVI, p. 313.
4. Quoted in Campbell, *Lanarkshire Miners*, p. 104; from Report of the Mining Commissioner, 1844, XVI, p. 20
5. Campbell, *Lanarkshire Miners*, p. 104.
6. A. Alison, 'Social and Moral Condition of the Manufacturing Districts of Scotland', in *Blackwood's Edinburgh Magazine*, 50 (Nov 1841).

7. 1842 Report, XVI, p. 362.

8. J.A. Hassan, *The Development of the Coal Industry in Mid and West Lothian 1815–1873* (Strathclyde University Ph.D. thesis); and J.H. McKay, *A Social History of the Scottish Shale Mining Community* (Open University Ph.D thesis, 1984).

9. Smout, 'Lead Mining in Scotland 1650–1850', in P.L. Payne (ed.), *Studies in Scottish Business Industry* (London, 1967).

10. 1844 Report; and 1845 Report XXVII.197

11. L. King, *Sair, Sair Wark* (Kelty, 2001), Chapter 5; and A.V. John, *By the Sweat of Their Brow: Women Workers at Victorian Coal Mines* (London, 1980), p. 57.

12. King, *Sair, Sair Wark*; and John, *By the Sweat of Their Brow*, pp. 50–9 for coverage of the aftermath of the Mines Act of 1842.

13. 1844 Report, p. 6.

14. 1844 Report, p. 6.

15. F. Reid, *Keir Hardie* (London, 1978), pp. 17–18.

16. 1844 Report, p. 14.

17. 1842 Report, XVI, p. 327.

18. Campbell, *Lanarkshire Miners*, especially chapter 7.

19. R.H. Campbell and J.B.A. Dow, *Source Book of Scottish Economic and Social History* (Oxford, 1968), p. 8.

20. J.A. Hassan, 'The Landed Estate, Paternalism, and the Coal Industry in Midlothian 1800–1880', in *Scottish Historical Review,* 59 (1980).

21. G.M. Wilson, *The Miners of the West of Scotland and their Trade Unions 1842–1874* (University of Glasgow Ph.D thesis, 1977), Chapter 2, for coverage of mining hazards.

22. R. Duncan, *Wishaw: Life and Labour in a Lanarkshire Mining Community 1790–1914* (Motherwell, 1986), chapter 4.

23. R.P. Arnot, *A History of the Scottish Miners* (London, 1955), p. 61.

24. McKay, *Shale Mining.*

25. Wilson, *Miners of the West of Scotland*, p. 59.

26. Wilson, *Alexander McDonald*, Chapter 3.

27. Smout, Lead Mining.

28. Duncan, *Wishaw*, p. 112; also B. E. Paterson, 'The Social and Working Conditions of the Ayrshire Mining Population', in *Ayrshire Collections,* 10 (1972), for Dalmellington.

29. Hassan, 'Landed Estate'.

30. Campbell, *Lanarkshire Miners*, Chapter 8.

31. Reid, *Keir Hardie*, p. 38.

32. Wilson, *Alexander McDonald*, Chapter 3.

33. W.S. Harvey, 'The Strike at Leadhills Mines, 1836', in *Local Historian* 17:1 (Feb 1986).

34. G.M. Wilson, 'The Strike Policy of the Miners of the West of Scotland 1842–1874', in I. MacDougall (ed.), *Essays in Scottish Labour History* (Edinburgh, 1979), p. 36. The 1842 episode is covered in the main texts, notably Wilson, *Alexander McDonald*, and Campbell, *Lanarkshire Miners*; also in a short study, R. Duncan, *Conflict and Crisis: Monklands Miners and General Strike 1842* (Monklands District Libraries, 1984).

35. Hassan, 'Landed Estate'.

36. Hassan, 'Landed Estate'; also Hassan, *Development* Chapter 9.

37. Reid, *Hardie,* p. 59.

38. Campbell, *Lanarkshire Miners*, p. 140.

39. Campbell, *Lanarkshire Miners* is the major source on Free Colliers and on sectarian rivalry in mining communities.

40. Duncan, *Wishaw*, Chapter 5; and R. Duncan, *Steelopolis: The Making of Motherwell 1750–1939* (Motherwell, 1990), Chapter 6

41. Campbell, *Lanarkshire Miners*, p. 162.

Chapter 5

1. The principal contextual source for this first section and for this chapter is A.B. Campbell, *The Scottish Miners 1874–1939, vol. 1: Industry, Work and Community* (Aldershot, 2000).
2. A. Slaven, *The Development of the West of Scotland: 1750–1960* (London, 1975), p. 168.
3. R.H. Campbell, *The Rise and Fall of Scottish Industry 1707–1939* (Edinburgh, 1980), Chapter 5.
4. R. Church, *The History of the British Coal Industry, vol. 3, 1830–1913* (Oxford, 1986), p. 400.
5. Quoted in Campbell, *Scottish Miners*, p. 61.
6. J. H. McKay, *A Social History of the Scottish Shale Mining Community* (Open University Ph.D thesis, 1984); and his major contribution in S. Cavanagh (ed.), *Pumpherston. The Story of a Shale Oil Village* (Edinburgh, 2002).
7. G. Anderson, *Down the Mine at Twelve* (Hamilton, 1985).
8. J. Hamilton (ed.), *Lanarkshire Coalminers and their Wives. Reminiscences from Coalburn and Surrounding Villages* (Hamilton, 2003), p. 58.
9. Hamilton, *Lanarkshire Coalminers*, p. 50.
10. Hamilton, *Lanarkshire Coalminers*, p. 43
11. Quoted in Campbell, *Scottish Miners*, pp. 79–80.
12. J. Archibald Henderson, *Autobiography*, p. 150; unpublished typescript (no date), copy in Motherwell Heritage Centre.
13. I. MacDougall (ed.), *Militant Miners* (Edinburgh, 1981) includes the recollections of John McArthur (pp. 1–167). The quotes in this section are from pp. 6–8.
14. F. Reid, *Keir Hardie* (London, 1978), p. 67.
15. Campbell, *Scottish Miners*, p. 97.
16. F. McLauchlan, '"Polish Labour" in the Scottish mines. From the miner's point of view', in *Economic Journal*, 17 (June 1907).
17. K. Lunn, 'Reactions and Responses: Lithuanian and Polish Immigrants in the Lanarkshire Coalfield 1880–1914', in *Scottish Labour History Journal*, 13 (1979).
18. Anderson, *Down the Mine at Twelve*, p. 43.
19. J. Anderson, *Coal! A History of the Coal Mining Industry in Scotland with special reference to the Cambuslang district* (Cambuslang Miners' Welfare Committee, 1943), pp. 31, 61.
20. I. MacDougall (ed.) *Voices from Work and Home* (Edinburgh, 2000), p. 106.
21. D. Kerr, *Shale Oil: Scotland: The World's Pioneering Oil Industry*, 2nd edn, 1999. pp. 32–4; McKay, *Social History of Scottish Shale Mining*, p. 557.
22. L. King, *Sair, Sair, Wark* (Kelty, 2001) for essential background.
23. Hamilton, *Lanarkshire Coalminers*, p. 138.
24. King, *Sair, Sair, Wark*, pp. 102–3.
25. MacDougall, *Voices from Work and Home*, pp. 74–5, and subsequent quotes in this section.
26. K. Durland, *Among the Fife Miners* (London, 1904).
27. Durland, *Among the Fife Miners*, p. 185.
28. Durland, *Among the Fife Miners*, p. 184.
29. Durland, *Among the Fife Miners*, p. 118.
30. J. Milligan, *The Memoirs of John Milligan of Dreghorn* (n.p., 1975), pp. 11–12.
31. Hamilton, *Lanarkshire Coalminers*, p. 71.
32. A. Moffat, *My Life with the Miners* (London, 1965), p. 40.
33. Royal Commission on the Housing of the Industrial Population of Scotland. Report. Parliamentary Papers. Command 8731 (1917), p. 137.

34. Royal Commission Report, p. 127.

35. Campbell, *Scottish Miners*, p. 221.

36. Royal Commission Report, p. 152.

37. R. Duncan, *Wishaw: Life and Labour in a Lanarkshire Mining Community 1790–1914* (Motherwell, 1986), Chapter 7; R. Duncan, *Steelopolis: The Making of Motherwell 1750–1939* (Motherwell, 1990), Chapter 5.

38. The evidence submitted by Ayrshire Miners' Union to the Royal Commission on Housing was published as a separate booklet, 'Ayrshire Miners' Rows 1913', in *Ayrshire Collections* 13:1 (1979).

39. J. Wheatley, *Miners, Mines, and Misery* (Glasgow, 1909).

40. Moffat, *My Life*, p. 12.

41. J.L. Carvel, *The Coltness Iron Company* (Edinburgh, 1948), p. 151.

42. A.B. Campbell, *The Scottish Miners 1874–1939 Volume 2: Trade Unions and Politics* (Aldershot, 2000): Chapters 1–3 provide essential context and information.

43. I. MacDougall (ed.), *Mungo Mackay and the Green Table: Newtongrange Miners Remember* (East Linton, 1995).

44. C. McMaster, 'The Gothenburg Experiment in Scotland', in *Scottish Labour History Review*, 3 (Glasgow, 1989).

45. R. Duncan, *Bothwellhaugh: A Lanarkshire Mining Community 1884–1965* (Glasgow, 1986).

46. W. S. Harvey, 'The Progress of Trade Unionism at the Leadhills Mines: 1836 to 1914' in *British Mining*, 39 (1989), pp. 47–52.

47. Campbell, *Scottish Miners, vol. 1*, pp. 39–40; see also discussion of sectarian division in his Chapter 7.

48. McKay, *Social History of the Scottish Shale Mining*, Chapter 15; also R. Ross, 'A Century of Shale', in *Cencrastus* (summer 1994).

49. Campbell, *Scottish Miners, vol. 1*, pp. 276–80.

50. R.P. Arnot, *A History of the Scottish Miners* (London, 1955): Chapters 4–6 cover the 1894 and 1912 disputes in some detail, as does Campbell.

51. Campbell, *Scottish Miners, vol. 1*, especially pp. 300–303; Duncan, *Steelopolis*, chapter 6.

52. Campbell, *Scottish Miners, vol. 2*, Chapters 2 and 3; John McArthur's detailed memoirs in *Militant Miners*, and an article by J.D. MacDougall, 'The Scottish Coalminer', in *The Nineteenth Century* (Dec 1927) are perceptive accounts by principal activists.

Chapter 6

Key secondary sources for this overview and chapter are: A.B. Campbell, *The Scottish Miners 1874–1939, vols 1 & 2* (Aldershot, 2000) and B. Supple, *History of the British Coal Industry, vol. 4, 1913–1946* (Oxford, 1987); while R. P. Arnot, *A History of the Scottish Miners* (London, 1955) still provides useful coverage.

1. R. Duncan, *Shotts Miners* (Motherwell, 1982), p. 3.

2. R. Duncan, *Steelopolis: The Making of Motherwell 1750–1939* (Motherwell, 1990), Chapter 9.

3. Scottish Home Department, *Scottish Coalfields, The Report of the Scottish Coalfields Committee*, PP (1944–45) (Command 6575), p. 144.

4. 1944 Report, p. 144.

5. 1944 Report, p. 119.

6. J. Hutchison, *Weavers, Miners and the Open Book. A History of Kilsyth* (1986), p. 190.

7. J. L. Carvel, *The Coltness Iron Company* (Edinburgh, 1948), pp. 178–9; 1944 Report, pp. 133–5.

8. *Miners' Welfare in Wartime, Report of the Miners Welfare Commission to June 1946* (1947).

9. A. Moffat, *My Life with the Miners* (London, 1965), pp. 85–6.

10. L. Mackenney (ed.), *Joe Corrie, Plays, Poems, and Theatre Writings* (Edinburgh, 1985), p. 138.

11. R. Smith, *Seven Steps in the Dark* (Barr, 1991), pp. 94–5.

12. Interview with the author, research for *Bothwellhaugh*.

13. Smith, *Seven Steps*, pp. 88–9, 94–5, 103.

14. I. MacDougall (ed.), *Voices from Work and Home* (Edinburgh, 2000), pp. 115–16.

15. MacDougall, *Voices*, p. 90.

16. Smith, *Seven Steps*, p. 119, and Chapter 3 on Bothwell.

17. A. McIvor and R. Johnston, 'Voices from the Pits. Health and Safety in Scottish Coal Mining since 1945', in *Scottish Economic and Social History* 22:2 (2002).

18. *Miners' Welfare in Wartime* pp. 47–8; G. Hutton, *Lanarkshire's Mining Legacy* (Cumnock, 1997), pp. 19–21.

19. Attributed to Bob Young, Bothwellhaugh.

20. I. MacDougall (ed.), *Militant Miners* (Edinburgh, 1981), p. 50.

21. Campbell, vol. 2; and S. Macintyre, *Little Moscows. Communism and Working Class Militancy in Inter-war Britain* (London, 1980) provide the essential context for Fife. Macintyre has a case study of Lumphinnans.

22. I. MacDougall (ed.), *Mungo Mackay and the Green Table: Newtongrange Miners Remember* (East Linton, 1995), pp. 63–4.

23. J.L. Carvel, *The New Cumnock Coalfield: A Record of its Development and Activities* (1946), p. 109.

24. Duncan, *Shotts Miners*, p. 7.

25. MacDougall, *Militant Miners*, p. 51.

26. Hutchison, A (ed.), *Corrie and Cardenden* (Workers' Educational Association, Edinburgh 1986), p 9

27. Docherty, M, *A Miner's Lass,* (Cowdenbeath, 1992), pp. 34–5

28. Duncan, *Shotts Miners*, pp. 8–9.

29. There are many studies of the General Strike and lockout; but see Campbell, vol. 2, Chapter 5 for Scotland, and especially MacDougall *Militant Miners* for Fife, including the full, vivid memoir of John McArthur and the letters of David Proudfoot, pp. 258–323, covering 1926. McIlroy, J., Campbell, A., and Gildart, K. (eds) *Industrial Politics and the 1926 Mining Lockout* (Cardiff, 2004) provides a new interpretation, including a chapter by Campbell on Scotland.

30. MacDougall, *Militant Miners*, p. 99.

31. Docherty, *A Miner's Lass*, p. 50.

32. Moffat, *My Life*, p. 34.

33. Duncan, *Shotts Miners*, p. 15.

34. Duncan, *Shotts Miners*, p. 14.

35. I. MacDougall (ed.), *Voices from the Hunger Marches, vol. II. Personal Recollections by Scottish Hunger Marchers of the 1920s and 1930s* (Edinburgh, 1991), p. 33.

36. A. Findlay, *Shale Voices* (Edinburgh, 1999), pp. 95–7. This is a delightful book and an essential source for the shale-mining communities of West Lothian.

37. Macintyre, *Little Moscows*, p. 62.

38. J. D. MacDougall, 'The Scottish Coalminer' (1927).

39. Moffat, *My Life*, pp. 19–24.

40. See especially the testimony of John McArthur, in MacDougall *Militant Miners;* also I. Macdougall (ed.), *Voices from the Hunger Marches, vol. 1* (Edinburgh, 1990) and vol. 2. Both books include important reminiscences of mining activists, notably Rab Smith (Fife), in vol. 1, pp. 83–110.

41. Campbell, vol. 1, chapter 7, and vol. 2, pp. 190–1.

42. Campbell, vol. 1, pp. 140–51, for analysis of strike record.

43. J. McCormack (with S. Pirani), *Polmaise. The Fight for a Pit* (1989), pp. 1–2.

44. This extraordinary 'hidden history' is revealed in a two-part specialist article, J. McIlroy and A. Campbell, 'Beyond Betteshanger: Order 1305 in the Scottish Coalfields during the Second World War', in *Historical Studies in Industrial Relations,* 15 (spring 2003); and 'Beyond Betteshanger: Order 1305 in the Scottish Coalfields during the Second World War: the Cardowan Story,' in *Historical Studies in Industrial Relations,* 16 (Autumn 2003).

Chapter 7

1. National Coal Board, *A Short History of the Scottish Coal Mining Industry* (Edinburgh, 1958), Chapter 6; and *Scotland's Coal Plan* (NCB, 1965). R.S. Halliday, *The Disappearing Scottish Colliery* (Edinburgh, 1990) is a useful analysis of the industry between the 1940s and 1980s.

2. P. Payne, 'The Decline of the Scottish Heavy Industries 1945–1983', in R. Saville (ed.), *The Economic Development of Modern Scotland 1950–1983* (Edinburgh, 1985), pp. 83–93; and W. Ashworth, *The History of the British Coal Industry,* vol. 5: 1946–1982 (Oxford, 1986) discuss these developments.

3. Halliday, *The Disappearing Scottish Colliery,* p. 14; and a two-part article, J. McNeil, 'The Fife Coal Industry 1947–1967. A Study of Changing Trends and their Implications', in *Scottish Geographical Magazine* (1973).

4. R. Smith, *Seven Steps in the Dark* (Barr, 1991), p. 196.

5. A. Moffat, *My Life with the Miners* (London, 1965), Chapter 14; R. Ross, 'A Century of Shale', in *Cencrastus* (summer 1994) for Burngrange.

6. *Scottish Miner,* January 1968.

7. I. Terris, *Twenty Years Down The Mines* (Ochiltree, 2001).

8. *Scottish Miner,* February 1973.

9. Smith, *Seven Steps,* pp. 224–5.

10. *Coalface* (Scottish Mining Museum) No 5, 1987; *Scottish Miner,* January 1968.

11. Ashworth, *British Coal Industry,* p. 561.

12. Smith, *Seven Steps,* pp. 221, 256–7.

13. A. McIvor and R. Johnston, 'Voices from the Pits. Health and Safety in Scottish Coal Mining since 1945', in *Scottish Economic and Social History* 22:2 (2002).

14. *Scotland's Coal Plan* (1955), p. 30.

15. H. Heughan, *Pit Closures at Shotts and the Migration of Miners* (Edinburgh, 1953).

16. W. Cullen, 'Redeployment of Miners in the Scottish Coal Industry 1947 to 1966' (unpublished typescript article, 1974, Scottish Mining Museum).

17. Moffat, *My Life,* Chapter 11.

18. Benarty Mining Heritage Group, *No More Bings in Benarty* (1992), p. 129.

19. J. McCormack (with S. Pirani), *Polmaise. The Fight for a Pit* (1989), p. 45.

20. J. Owens, *Miners 1984–1994. A Decade of Endurance* (Edinburgh, 1994), p. 22.

21. Contemporary accounts used here include: Owens, *Miners;* S. McGrail and V. Patterson, *For As Long As It Takes. Cowie Miners in the Strike 1984–85* (1985); McCormack, *Polmaise;* C. Levy and Mauchline Miners' Wives, *A Very Hard Year. The 1984–85 Miners Strike in*

Mauchline (Workers Educational Association, Glasgow, 1985); and *No More Bings in Benarty*. Alex Maxwell, a leading member of the Benarty Mining Heritage Group, published his version of the conflict and other local historical background in *Chicago Tumbles. Cowdenbeath and the Miners' Strike* (1994). L. King, *Sair, Sair, Wark* (Kelty, 2001) has a useful chapter on women's support groups.

22. Owens, *Miners*, pp. 6–7.
23. *A Very Hard Year*, pp. 16–17.
24. *A Very Hard Year*, p. 17.
25. V. Seddon (ed.), *The Cutting Edge. Women and the Pit Strike* (1986), pp. 224–5, contribution by Cath Cunningham.
26. Owens, *Miners*, p. 23, testimony of Helen Gray, Cumnock.
27. I. MacDougall (ed.) *Voices from Work and Home* (Edinburgh, 2000), pp. 145–6, testimony of Margaret Russell.
28. *A Very Hard Year*, p. 4.
29. Owens, *Miners*, p. 95.
30. McCormack, *Polmaise*.
31. For example, an in-depth investigative report in the *Guardian,* 12 August, 1985, which prompted a reappraisal of the police action.
32. *No More Bings in Benarty*, p. 130.
32. *Scottish Miner,* 1984–85 issues; also December 1987; June 1988.

Index

THE MINEWORKERS